AF411334

Omics Technologies in Food Science

Editors

Yelko Rodríguez-Carrasco
Bojan Šarkanj

MDPI • Basel • Beijing • Wuhan • Barcelona • Belgrade • Manchester • Tokyo • Cluj • Tianjin

Editors
Yelko Rodríguez-Carrasco
Food Science and Technology
University of Valencia
Valencia
Spain

Bojan Šarkanj
University of the North as the
Head of the Department of
Food Technology
University of the North
Koprivnica
Croatia

Editorial Office
MDPI
St. Alban-Anlage 66
4052 Basel, Switzerland

This is a reprint of articles from the Special Issue published online in the open access journal *Foods* (ISSN 2304-8158) (available at: www.mdpi.com/journal/foods/special_issues/omics_food).

For citation purposes, cite each article independently as indicated on the article page online and as indicated below:

LastName, A.A.; LastName, B.B.; LastName, C.C. Article Title. *Journal Name* **Year**, *Volume Number*, Page Range.

ISBN 978-3-0365-4612-4 (Hbk)
ISBN 978-3-0365-4611-7 (PDF)

Contents

About the Editors

Yelko Rodríguez-Carrasco

Dr. Yelko Rodríguez Carrasco is Professor at the Preventive Medicine and Public Health, Food Science, Toxicology and Forensic Medicine Department of University of Valencia (UV, Spain). His research and teaching revolve around food science and technology and food toxicology areas, publishing more than 50 scientific articles, most of them in Q1 journals. Professor Rodríguez has participated in several regional, national, and international research projects, and has contributed in more than 60 scientific congresses. On a professional level, he has worked for a multinational leader in the agri-food sector in Germany and Brazil. Professor Rodríguez graduated in both Human Nutrition and Dietetics and Food Science and Technology, and holds a Master's degree in Food Quality and Safety achieved at the University of Valencia, for which he received an extraordinary award. In 2015, he achieved his International PhD in Food Science at the same institution, obtaining the qualification cum laude and an extraordinary doctorate award. Dr. Yelko Rodríguez has conducted some predoctoral and postdoctoral research visits at HAW Hamburg University of Applied Sciences, BOKU University of Vienna, and at the University of Naples "Federico II". Professor Rodríguez is a member of the research group "RiskTox" (UV), and also member of the Spanish Association of Toxicology (AETOX).

Bojan Šarkanj

Dr. Bojan Šarkanj achieved his PhD in 2014. From October 2018 until today, he has been working at the University of the North as the Head of the Department of Food Technology. He is scientifically engaged in analytical and molecular mycotoxicology. Thus far, he has published more than 50 scientific papers. He has been the leader of five scientific projects, and as a collaborator he worked on fifteen other projects. He is a member of the Croatian Toxicological Society, the Croatian Microbiological Society, the Croatian Society for Biochemistry and Molecular Biology, and the Global Harmonization Initiative (GHI), section for mycotoxins.

Preface to "Omics Technologies in Food Science"

Food science has greatly developed over the last few years. This has led to increased attention towards the origin and quality of raw materials, as well as their derived food products. The continuous advance in molecular biology has allowed the implementation of efficient and universal omics tools to unequivocally identify the origin of food items and their traceability. This Book presents research papers in the fields of omics and multi-omics for food science applications. This Book approaches aspects such as food safety and food quality, as well as new technologies including metabolomics, metagenomics, nutrigenomics, transcriptomics, and proteomics. Obtaining and discussing new information is an important step toward continuous research in the area of food science.

Yelko Rodríguez-Carrasco and Bojan Šarkanj

Editors

Editorial

Foodomics: Current and Future Perspectives in Food Analysis

Yelko Rodríguez-Carrasco 

Department of Preventive Medicine and Public Health, Food Science, Toxicology and Forensic Medicine, Faculty of Pharmacy, University of Valencia, Burjassot, 46100 Valencia, Spain; yelko.rodriguez@uv.es; Tel.: +34-963544228

Citation: Rodríguez-Carrasco, Y. Foodomics: Current and Future Perspectives in Food Analysis. *Foods* **2022**, *11*, 1238. https://doi.org/10.3390/foods11091238

Received: 18 April 2022
Accepted: 20 April 2022
Published: 26 April 2022

Publisher's Note: MDPI stays neutral with regard to jurisdictional claims in published maps and institutional affiliations.

Climate change, an increase in population, and the recent pandemic crisis triggered by SARS-CoV-2 have all contributed to a period of global problems. Due to increased food traffic and the number of prospective customers, it is vital to ensure food safety in terms of quality, traceability, and safety in this situation. Furthermore, environmental stressors such as meteorological disasters, rising global temperatures, and rising CO_2 levels are on the rise and reducing crop yields. Moreover, there is an increasing interest in the diet–health dyad, as well as the link between certain chronic non-communicable diseases and diet. All of these new demands and problems are difficult to meet using traditional analytical chemistry in the food sciences.

Foodomics was conceptualized to address the issues of food science. It is characterized as a discipline that explores the fields of food and nutrition through the application of omics technologies. This discipline employs high-resolution techniques such as mass spectrometry, nuclear magnetic resonance, and next-generation sequencing, as well as a variety of mathematical and computer tools to handle and interpret data.

As a result, high-resolution analytical techniques are gradually being introduced into the food sciences, along with instruments that allow for the interpretation of the huge amounts of data generated. Because of their immense utility and development potential, considerable emphasis is currently made on the development of resolution power, analysis speed, affordability, and downsizing of these instruments due to the growth in food safety standards.

Food safety is a critical component of the food production and distribution process. Toxicological analyses allow for the qualitative and quantitative detection of substances and organisms that have the potential to affect people's health (food hazards) or that can signal their presence directly or indirectly. Because of the vast quantity of information that can be collected from a single examination, foodomics allows for the adoption of approaches capable of detecting and evaluating new food hazards, as well as better efficiency in toxicological research. Foodomics can also help to understand how the production chain's operations affect food, allowing for the development of optimization, assessment, and monitoring strategies. On the other hand, adulterations and fraud, both unintentional and intentional for financial benefit, are a serious problem that needs to be addressed. These actions cost the food business some billion euros and jeopardize public health because adulterated foods may contain ingredients that are capable of triggering diseases in customers who are ignorant of their presence. Hence, food fraud is also being investigated and controlled throughout omics technologies.

This Special Issue intends to contribute to the growth of this discipline in order to highlight its prospective uses and how it fits into the current scientific landscape in the area of food analysis.

Martínez-Alonso et al. (2022) [1] demonstrated that red beans are a rich source of bioactive substances such as flavanols and anthocyanidins, which offer antiradical activity and human health advantages, according to the chemical profile obtained from this study. In their work, red bean extract (diluted 1:8) demonstrated a significant increase in human liver cancer cell line (HepG2) proliferation after 24 h of exposure. Additionally, diluted red bean extract resulted in a reduction in reactive oxygen species (ROS) generation compared to the control at 120 min. The latter result was related to antioxidant activity, implying that the red bean extract could modulate oxidative stress in HepG2 cells.

Shi et al. (2022) [2] revealed the complimentary patterns of nutrient accumulation in distinct species and decoded the species-specific patterns of bioactive chemicals in three key staple crops and three fruits. Sweet corn had the most vitamins and amino acids of the three crops, but rice and wheat were vitamin and amino acid deficient. Mango was the most vitamin- and amino-acid-rich of the three fruits. Crops were high in fats compared to fruits. Overall, this research presented metabolomic evidence for a healthy diet, emphasizing the importance of macronutrients and micronutrients.

Lee et al. (2021) [3] used untargeted metabolite profiling methodologies to conduct a complete metabolite profiling of various soy fermented products and raw soybeans. It was reported in this research that variations in the metabolomes and volatolomes of soy products may be altered by changing processing steps, microbial successions, and fermentation time.

Keawkim et al. (2021) [4] highlighted that combined metabolomics and flavoromics could be utilized to identify the major metabolites and flavor components that vary in sacha inchi seeds during germination with the aim of establishing the optimum nutritional values and guide the improvement of pharmaceutical products or functional meals manufactured from sacha inchi seeds. In this research, it was reported that germinated sacha inchi seeds have higher levels of amino acids, organic acids, total phenolic compounds, and antioxidant activity than ungerminated seeds.

Uršulin-Trstenjak et al. (2021) [5] applied foodomics as a tool for the development of healthy food that is tailored to individual health issues and the prevention of food-related disorders in order to promote human longevity. Diet difficulties, which often lead to malnutrition, are common among the elderly in nursing facilities. To avoid nutritional danger, this study emphasizes the significance of examining and regularly monitoring the nutritional health of persons in elderly homes.

Rocchetti et al. (2021) [6] evaluated the potential of a metabolomics-based technique combined with retrospective high-resolution mass spectrometry (HRMS) screening to evaluate the mycotoxin profile of bulk milk samples. Overall, the HRMS technique putatively identified 46 mycotoxins. Zearalenol, mycophenolic acid, tentoxin, and apicidin were revealed to be among the most discriminant markers in this study. Although further confirmation studies were required, this work suggested potential carry-over and metabolization phenomena of mycotoxins in milk.

Cai et al. (2021) [7] reported that long-term 1-methylcyclopropene (1-MCP) treatment produces papaya ripening disorder, while appropriate 1-MCP treatment significantly prevents the ripening of papaya fruit. Their findings suggested that by targeting the ethylene and auxin signaling pathways, the miRNAs may play a significant role in regulating fruit ripening. Unsuitable 1-MCP therapy can cause fruit ripening disease by disrupting miRNA function.

The study of Kafantaris et al. (2021) [8] was the first to use a global transcriptome method to evaluate pine honey's antibacterial effects and mechanism of action. Pine honey had a substantial effect on the transcriptomic profile of *Pseudomonas aeruginosa*. This study demonstrated that pine honey had a suppressive effect on *P. aeruginosa* genome expression, with more genes down-regulated than up-regulated. These findings could help to unravel the molecular pathways and biological processes involved in pine honey's antibacterial action, which could aid in treating and control *P. aeruginosa* infection and pathogenicity.

Yu et al. (2021) [9] found that age had a substantial impact on meat quality parameters and meat exudate metabolome profiles. Metabolites linked to ATP synthesis and metabolism (creatine and hypoxanthine), antioxidation (GSSG and carnosine), and proteolysis (dipeptides and tripeptides) could be used as biomarkers to track aging times and detect changes in meat quality, such as increased lipid and protein oxidation, discoloration, and tenderness.

Based on the above-mentioned studies, the assurance of food quality and safety will be the foremost need in the future of food sciences. In the coming years, the worldwide circulation of food and related raw materials will expand, resulting in an increase in contamination. Furthermore, numerous processed items created with various ingredients from around the world will share storage rooms and production lines, making food safety incidents more difficult to regulate. As a result, assuring food safety, quality, and traceability will be considerably more difficult and critical than it is now. Foodomics is currently demonstrating its potential to respond to the demands of food science, even though it will take more time to develop, consolidate, and gain recognition. Based on the manuscripts listed in this Special Issue and on the others found in the literature, foodomics has a wide range of experimental methodologies and applications in the areas of food safety, quality control, authenticity, food process design, biotechnology, and nutrition. The existence of these current and future challenges, which are still far from being fully resolved, highlights the need to continue working on these issues in order to acquire a comprehensive perspective on the complicated situation in food and nutrition sciences.

Funding: This research was funded by the Spanish Ministry of Science and Innovation Project (PID2020-115871RB-I00).

Data Availability Statement: No new data were created or analyzed in this study. Data sharing is not applicable to this article.

Conflicts of Interest: The author declares no conflict of interest.

References

1. Martínez-Alonso, C.; Taroncher, M.; Castaldo, L.; Izzo, L.; Rodríguez-Carrasco, Y.; Ritieni, A.; Ruiz, M.-J. Effect of Phenolic Extract from Red Beans (*Phaseolus vulgaris* L.) on T-2 Toxin-Induced Cytotoxicity in HepG2 Cells. *Foods* **2022**, *11*, 1033. [CrossRef] [PubMed]
2. Shi, Y.; Guo, Y.; Wang, Y.; Li, M.; Li, K.; Liu, X.; Fang, C.; Luo, J. Metabolomic Analysis Reveals Nutritional Diversity among Three Staple Crops and Three Fruits. *Foods* **2022**, *11*, 550. [CrossRef] [PubMed]
3. Lee, S.-H.; Lee, S.; Lee, S.-H.; Kim, H.-J.; Singh, D.; Lee, C.-H. Integrated Metabolomics and Volatolomics for Comparative Evaluation of Fermented Soy Products. *Foods* **2021**, *10*, 2516. [CrossRef] [PubMed]
4. Keawkim, K.; Lorjaroenphon, Y.; Vangnai, K.; Jom, K.N. Metabolite—Flavor Profile, Phenolic Content, and Antioxidant Activity Changes in Sacha Inchi (*Plukenetia volubilis* L.) Seeds during Germination. *Foods* **2021**, *10*, 2476. [CrossRef] [PubMed]
5. Uršulin-Trstenjak, N.; Šarkanj, I.D.; Sajko, M.; Vitez, D.; Živoder, I. Determination of the Personal Nutritional Status of Elderly Populations Based on Basic Foodomics Elements. *Foods* **2021**, *10*, 2391. [CrossRef] [PubMed]
6. Rocchetti, G.; Ghilardelli, F.; Masoero, F.; Gallo, A. Screening of Regulated and Emerging Mycotoxins in Bulk Milk Samples by High-Resolution Mass Spectrometry. *Foods* **2021**, *10*, 2025. [CrossRef] [PubMed]
7. Cai, J.; Wu, Z.; Hao, Y.; Liu, Y.; Song, Z.; Chen, W.; Li, X.; Zhu, X. Small RNAs, Degradome, and Transcriptome Sequencing Provide Insights into Papaya Fruit Ripening Regulated by 1-MCP. *Foods* **2021**, *10*, 1643. [CrossRef] [PubMed]
8. Kafantaris, I.; Tsadila, C.; Nikolaidis, M.; Tsavea, E.; Dimitriou, T.G.; Iliopoulos, I.; Amoutzias, G.D.; Mossialos, D. Transcriptomic Analysis of *Pseudomonas aeruginosa* Response to Pine Honey via RNA Sequencing Indicates Multiple Mechanisms of Antibacterial Activity. *Foods* **2021**, *10*, 936. [CrossRef] [PubMed]
9. Yu, Q.; Cooper, B.; Sobreira, T.; Kim, Y.H.B. Utilizing Pork Exudate Metabolomics to Reveal the Impact of Aging on Meat Quality. *Foods* **2021**, *10*, 668. [CrossRef] [PubMed]

Article

Effect of Phenolic Extract from Red Beans (*Phaseolus vulgaris* L.) on T-2 Toxin-Induced Cytotoxicity in HepG2 Cells

Carmen Martínez-Alonso [1,†], Mercedes Taroncher [1,†], Luigi Castaldo [2], Luana Izzo [2], Yelko Rodríguez-Carrasco [1,*], Alberto Ritieni [2] and María-José Ruiz [1]

1 Department of Preventive Medicine and Public Health, Food Science, Toxicology and Forensic Medicine, Faculty of Pharmacy, University of Valencia, Burjassot, 46100 Valencia, Spain; carmar36@alumni.uv.es (C.M.-A.); mercedes.taroncher@uv.es (M.T.); m.jose.ruiz@uv.es (M.-J.R.)
2 Department of Pharmacy, Faculty of Pharmacy, University of Naples "Federico II", Via Domenico Montesano 49, 80131 Naples, Italy; luigi.castaldo2@unina.it (L.C.); luana.izzo@unina.it (L.I.); alberto.ritieni@unina.it (A.R.)
* Correspondence: yelko.rodriguez@uv.es
† These authors contributed equally to this work.

Abstract: Red beans contain human bioactive compounds such as polyphenols. Several in vitro studies have proposed the natural compounds as an innovative strategy to modify the toxic effects produced by mycotoxins. Hence, in this work, a complete investigation of the polyphenolic fraction of red beans was performed using a Q-Orbitrap high-resolution mass spectrometry analysis. Notably, epicatechin and delphinidin were the most detected polyphenols found in red bean extracts (3.297 and 3.108 mg/Kg, respectively). Moreover, the red bean extract was evaluated against the T-2 toxin (T-2) induced cytotoxicity in hepatocarcinoma cells (HepG2) by direct treatment, simultaneous treatment, and pre-treatment assays. These data showed that T-2 affected the cell viability in a dose-dependent manner, as well as observing a cytotoxic effect and a significant increase in ROS production at 30 nM. The simultaneous treatment and the pre-treatment of HepG2 cells with red bean extract was not able to modify the cytotoxic T-2 effect. However, the simultaneous treatment of T-2 at 7.5 nM with the red bean extract showed a significant decrease in ROS production, with respect to the control. These results suggest that the red bean extract could modulate oxidative stress on HepG2 cells.

Keywords: beans; phenolic compounds; T-2 toxin; HepG2 cells; reactive oxygen species

Citation: Martínez-Alonso, C.; Taroncher, M.; Castaldo, L.; Izzo, L.; Rodríguez-Carrasco, Y.; Ritieni, A.; Ruiz, M.-J. Effect of Phenolic Extract from Red Beans (*Phaseolus vulgaris* L.) on T-2 Toxin-Induced Cytotoxicity in HepG2 Cells. *Foods* **2022**, *11*, 1033. https://doi.org/10.3390/foods 11071033

Academic Editor: Paula C. Castilho

Received: 21 March 2022
Accepted: 28 March 2022
Published: 2 April 2022

Publisher's Note: MDPI stays neutral with regard to jurisdictional claims in published maps and institutional affiliations.

1. Introduction

Red beans (*Phaseolus vulgaris* L.), in addition to being a great source of vegetable protein, fiber, and certain micronutrients in the human diet, contain a great variety of bioactive compounds. Bioactive compounds are simple substances that have biological activity, associated with their ability to modulate one or more metabolic processes, which results in the promotion of better health conditions. Different bioactive compounds have been studied for their positive effect on human health, such as the following: enzymes, probiotics, prebiotics, fibers, phytosterols, peptides, proteins, saponins, unsatured fatty acids, and phenolic compounds, among others [1]. Particularly, colored beans have significant amounts of phenolic compounds. Phenolic compounds are classified as flavonoids (flavones, flavonols, flavanones, isoflavones, anthocyanins, chalcones, dihydrochalcones, and catechins), phenolic acids (hydroxybenzoic, hydroxyphenyl acetic, hydroxyphenyl pentanoic, and cinnamic hydroxyl acids), tannins, stilbenes, and lignans. Phenolic acids, flavonoids, and anthocyanidins are the main phenolic compounds identified and characterized in beans [2,3]. Phenolic compounds determine the color of the seeds of these legumes; hence, in general, a higher phenolic content is observed in more pigmented beans [4]. Phenolic compounds, besides from contributing to the smell, taste, and color of food, have a long-term intake that could play a bioactive role due to their antioxidant activity, which has been related to the prevention of obesity, cardiovascular and neurodegenerative diseases, cancer, and diabetes, as well as exhibiting anti-inflammatory, antimutagenic, and antibacterial properties [5,6]. Madhujith et al. demonstrated that beans, especially those with colored skins, possess strong antioxidant activity as measured by different model systems [7]. Epidemiological studies correlated the consumption of procyanidin-rich foods with a lower incidence of inflammatory disease and diseases of multifactorial pathogenesis [8]. Similarly, the transcription and secretion of proinflammatory cytokines, including IL-1β, IL-2, IL-6, TNF-α, and interferon-γ, could be down-regulated by procyanidins, as reported in some in vitro and in vivo studies [9,10].

Despite the fact that legumes are protein rich foods, they are lacking in sulphur-containing amino acids. On the other hand, cereals contain sulphur amino acids but are limited in the essential amino acid lysine. Hence, a combination of legumes and cereals would improve the protein and nutrient density of the subsequent food products. However, the resulting high nutritional value food products could be susceptible to deterioration by fungal contamination, accompanied by the production of mycotoxins [11].

Mycotoxins are the secondary metabolites produced by filamentous fungi. The species assigned to the *Aspergillus*, *Penicillium*, *Alternaria*, *Claviceps*, and *Fusarium* genera produce a wide range of mycotoxins, which can contaminate food and feed, resulting in a significant threat to human and animal health [12]. Trichothecenes are a complex group of tetracyclic sesquiterpenoids produced by several *Fusarium* spp. Type A and B trichothecenes are the most relevant mycotoxins reported in food and feed regarding their incidence and concentration [13]. Among the trichothecenes, T-2 toxin (T-2) is the compound that shows the highest toxicity. Several studies have reported that T-2 toxin causes multiple damages to organs such as the kidney, liver, brain, gastrointestinal tract, and bone marrow [14]. Similarly, the toxic effects derived from repeated exposure to T-2 include genotoxicity, immunotoxicity, neurotoxicity, and reproductive toxicity [15,16]. Therefore, the European Food Safety Authority (EFSA) established a tolerable daily intake (TDI) for T-2 and its main metabolite HT-2 of 0.02 μg/Kg body weight (bw) to limit their exposure [13].

Based on the abovementioned information, the aims of this work are as follows: (i) to evaluate the total phenolic content (TPC), the phenolic profile, using ultra-high-performance liquid chromatography coupled with high-resolution mass spectrometry (UHPLC-Q-Orbitrap HRMS), and the antiradical activity of red bean extracts, and (ii) to assess the effect of the red bean extract on T-2 toxin-induced cytotoxicity in human hepatocarcinoma (HepG2) cells.

2. Materials and Methods

2.1. Chemicals and Reagents

Methanol (MeOH) and ethanol (EtOH) of HPLC grade were acquired from Merck (Darmstadt, Germany). Formic acid and ammonium formate were obtained from Fluka (Milan, Italy). Ethyl acetate was purchased from Merck Life Science S.L. (Madrid, Spain). Deionized water (resistivity < 18 MΩ cm) was obtained using a Milli-Q water purification system (Millipore, Bedford, MA, USA).

The chemical reagents and cell culture components used, namely Dulbecco's Modified Eagle's Medium (DMEM), penicillin, streptomycin, trypsin/EDTA solutions, phosphate buffered saline (PBS), Newborn Calf Serum (NBCS), methylthiazoltetrazolium salt (MTT) dye, dimethyl sulfoxide (DMSO), Sorensen's glycine buffer, dichlorodihydrofluorescein diacetate (H_2-DCFDA), 2,2-diphenyl-1-picrylhydrazyl (DPPH), Folin–Ciocalteu's phenol reagent, gallic acid ($C_7H_6O_5$), potassium chloride (KCl), sodium hydroxide (NaOH), hydrochloric acid (HCl), sodium chloride (NaCl), sodium phosphate dibasic (Na_2HPO_4), potassium phosphate monobasic (KH_2PO_4), and sodium carbonate (Na_2CO_3), were acquired from Sigma-Aldrich (Barcelona, Spain).

The standard of T-2 (MW: 466.52 g/mol) was purchased from Sigma-Aldrich (Barcelona, Spain). Standards of polyphenols (purity > 98%), namely protocatechuic acid, cyanidin 3,5-diglucoside, epicatechin, chlorogenic acid, cyanidin 3-galactoside, caffeic acid, catechin, *p*-cumaric acid, apigenin 7-glucoside, genistein, delphinidin, naringin, cyanidin, rosmarinic acid, myricitrin, diosmin, isoquercetin, rutin, kaempferol 3-glucoside, vitexin, ellagic acid, luteolin 7-glucoside, myricetin, diadzein, quercetin, delphinidin 3,5-diglucoside, naringenin, luteolin, kaempferol, and apigenin, were acquired from Sigma-Aldrich (Milan, Italy). Stock solutions of T-2 were prepared in MeOH at appropriate working concentrations and maintained in the dark at $-20\,^\circ$C.

2.2. Preparation of Red Bean Extract

The polyphenols were extracted from red beans according to the procedure reported in the literature with some modifications [17]. In summary, 0.5 g of ground beans was extracted with 10 mL of a solution MeOH:H_2O. The assayed mixtures were 20:80, 30:70, 50:50, 70:30, and 80:20 (*v/v*). Hydrochloric acid 2N was added to the sample until a pH adjustment to 2, to avoid the protonated forms of carboxylic groups present in polyphenols. Then, the mixture was subjected to a horizontal shaker (250 rpm) at room temperature for 3 h and centrifuged for 3 min at 3500 rpm. Finally, MeOH was evaporated from the acidified sample extract under reduced pressure (250 mbar) at 50 $^\circ$C for 15 min by Buchi Rotavapor R-200 (Buchi, Postfach, Switzerland) and the aqueous extract containing polyphenols was filtered with a 0.2 μm polytetrafluoroethylene (PTFE) filter and stored in amber glass flask at 4 $^\circ$C.

2.3. Determination of Total Phenolic Content (TPC)

The Folin–Ciocalteu assay was used to determine the total phenolic content in accordance with the procedure reported by Izzo et al. [18]. Briefly, 0.5 mL of red bean extract or blank (deionized water) was diluted with deionized water (4.5 mL) and 0.25 mL of Folin–Ciocalteu reagent 1 N was added. Then, 1 mL of 2% sodium carbonate solution was added and the mixture was allowed to stand at room temperature for 1 h in dark conditions. Finally, the absorbance was measured with a spectrophotometer at 765 nm against a reagent blank. The analysis was carried out in triplicate and the results were expressed as mg of gallic acid equivalents (GAE) per Kg of sample.

2.4. Determination of Polyphenolic Profile

The polyphenolic profile of the red bean extracts was carried out on a UHPLC-Q Exactive Orbitrap-HRMS system (Thermo Fisher Scientific, Waltham, MA, USA), composed of a Dionex Ultimate 3000 liquid chromatograph equipped with a solvent rack compartment (SRD-3x00), a quaternary rapid separation pump (LPG-3400RS), a rapid separation autosampler (WPS-3000RS), and a temperature-controlled column compartment (TCC-3000SD). The chromatographic separation was performed on a Kinetex F5 (50 × 2.1 mm; 1.7 μm) reverse-phase column (Phenomenex, Milan, Italy) at 25 $^\circ$C. The mobile phase consisted of water containing 0.1% formic acid (A) and MeOH containing 0.1% formic acid (B). The separation gradient consisted of an initial 0% of phase B, increasing to 40% B in 1 min, 80% B in a further 1 min, and 100% B in 3 min. Then, the gradient was held at 100% B for 4 min and reduced to 0% B in 2 min, followed by 2 min of column re-equilibration at 0% B. The total run time was 13 min. The flow rate was 0.5 mL/min and the injection volume was 1 μL.

The mass spectrometer was equipped with an electrospray (ESI) source that simultaneously operates in positive and negative ion switching mode. Full ion MS and all ion fragmentation (AIF) were set as scan events. The following settings were used in full MS mode: resolution power of 70,000 Full Width at Half Maximum (FWHM) (defined for m/z 200); automatic gain control (AGC) target: 1×10^6; scan range: 80–1200 m/z; injection time set to 200 ms; scan rate set at 2 scan/s. The ion source parameters were as follows: sheath gas pressure: 18; auxiliary gas: 3; spray voltage: 3.5 kV; capillary temperature: 320 $^\circ$C; S-lens RF level: 60; auxiliary gas heater temperature: 350 $^\circ$C. For the scan event of AIF, the parameters in the negative and positive mode were set as follows: mass resolving power = 17,500 FWHM; ACG target = 1×10^5; maximum injection time = 200 ms; scan time = 0.10 s; scan range = 80–1200 *m/z*; isolation window to 5.0 *m/z*; retention time to 30 s. The collision energy was varied between 10 and 60 eV to acquire representative product ion spectra.

For the identification and confirmation of the molecular ion and fragments, a mass tolerance below 5 ppm was set. Data analysis and processing were performed using Xcalibur software, v. 3.1.66.10 (Xcalibur, Thermo Fisher Scientific, Waltham, MA, USA).

2.5. Determination of Antiradical Activity (DPPH)

The total free radical scavenging activity of the red bean extracts was determined using the method reported in the literature with modifications [19]. Briefly, DPPH (4 mg) was solubilized in 10 mL of MeOH and then diluted to reach an absorbance value of 0.90 ($\pm$0.05) at 517 nm. This solution was used to perform the assay and 200 µL of red bean extract was added to 1 mL of working solution. The mixture was vortexed, kept for 90 min in the dark, and centrifuged for 5 min at 11,000 rpm. Finally, the decreased absorbance was measured at 517 nm. The analysis was carried out in triplicate and the results were expressed as mmol Trolox Equivalents (TE) per Kg of sample.

2.6. Cell Culture

Human hepatocarcinoma (HepG2) cells (ATCC: HB-8065) were cultured in DMEM medium supplemented with 10% NBCS, 100 U/mL penicillin, and 100 mg/mL strepto-mycin. The cells were maintained at pH 7.4, 5% CO_2 at 37 °C, and 95% air atmosphere at constant humidity. The cells were subcultured routinely twice a week with only a small number of sub-passages (<20 passages) in order to maintain genetic homogeneity. HepG2 cells were subcultured after trypsinization in a 1:3 split ratio. The medium was changed every 5 days. The final mycotoxin concentrations tested were achieved by adding T-2 mycotoxin to the culture medium, with a final MeOH concentration $\leq$ 1% (*v/v*).

2.7. HepG2 Cells Treatment

The HepG2 cells were cultured in 96-well tissue culture plates by adding 200 µL/well of density at 2×10^4 cells/well. After the cells reached 80% confluence, the culture medium was replaced with a fresh medium containing different concentrations of T-2 (7.5, 15, and 30 nM) and serial dilutions of red bean extracts (from 1:32 to 1) buffered to pH 7.4. Then, the plates were incubated in the dark at 37 °C and 5% CO_2 for 24 h. The assayed T-2 concentrations correspond to sublethal T-2 concentrations for HepG2 cells (<IC_{50}), based on previous studies carried out in our laboratory [20], and they were $IC_{50}/2$, $IC_{50}/4$, and $IC_{50}/8$, respectively.

The following two more assays were performed: simultaneous treatment and pre-treatment. On one hand, for pre-treatment studies, HepG2 cells were exposed to one red bean extract dilution according to previous cell proliferation assays (red bean extract 1:8 dilution) for 1 and 24 h. Then, the medium containing the red bean extract was removed and cells were exposed at the T-2 concentrations described above for 24 h. On the other hand, to conduct studies of simultaneous treatment, HepG2 cells were exposed to the assayed T-2 concentrations and the 1:8 diluted red bean extract for 24 h.

Appropriate controls containing the same amount of solvent were included in each experiment.

2.8. Determination of Cell Viability

Cell viability was determined in HepG2 cells by the MTT assay. The MTT assay is based on the capacity of viable cells to metabolize, via a mitochondrial-dependent reaction, specifically, the reduction of yellow tetrazolium salt to an insoluble purple formazan crystal. The MTT assay was carried out according to the procedure reported by Ruiz et al. [21]. In summary, after treatment studies, the medium containing the compounds was removed and each well received 200 µL of fresh medium containing 50 µL of MTT. The plates were returned to the incubator in the dark at 37 °C and 5% CO_2 for 3 h. Then, the MTT solution was removed and 200 µL of DMSO was added, followed by 25 µL of Sorensen's glycine buffer. The absorbance was measured at 620 nm on a Wallace Victor^2, model 1420 multilabel counter (PerkinElmer, Turku, Finland). The blank absorbance value (from wells without cells but treated with MTT) was subtracted from all absorbance values.

Cell viability was expressed as a percentage relative to control cells ($\leq$1% MeOH). Three independent experiments were conducted with eight replicates each, and the results were expressed as the mean $\pm$ standard error of the mean (SEM) of different independent experiments.

2.9. Determination of Reactive Oxygen Species (ROS)

Intracellular ROS production was monitored in HepG2 cells by adding H_2-DCFDA [22]. H_2-DCFDA is taken up by cells and then deacetylated by intracellular esterases; the resulting non-fluorescent 2′,7′-dichlorodihydrofluorescein (H_2-DCF) is converted to greatly fluorescent dichlorofluorescein (DCF) when oxidized by ROS. Briefly, 2×10^4 cells/well were seeded in a 96-well black polystyrene culture microplate. Once cells exhibited 80% confluence, the culture medium was replaced and cells were loaded with 20 µM H_2-DCFDA in a fresh medium for 20 min in darkness. Then, H_2-DCFDA was removed and replaced by a fresh medium containing 1:8 diluted red bean extract, T-2 (30, 15, and 7.5 nM), and the combination of diluted red bean extract with T-2 at the different concentrations assayed. Finally, the fluorescence emitted by the DCF was monitored at different times (0, 5, 15, 30, 45, 60, 90, and 120 min) on a Wallace Victor^2, model 1420 multilabel counter (PerkinElmer, Turku, Finland), at excitation/emission wavelengths of 485/535 nm, respectively. The blank absorbance value (from wells without cells but treated with H_2-DCFDA) was subtracted from all absorbance values.

The determinations were performed in three independent experiments with 24 replicates each and the results are expressed as an increase (%) in fluorescence in respect to control cells.

2.10. Statistical Analysis

Statistical analysis of data was carried out using the Statgraphics version 16.01.03 statistical package (IBM Corp., Armonk, NY, USA). Data are expressed as mean $\pm$ SEM of different independent experiments. The statistical analysis of the results was performed by Student's *t*-test for paired samples. The differences between groups were analyzed by employing one-way analysis of variance (ANOVA) continued by the Tukey HDS *post-hoc* test for multiple comparisons. Statistical significance was considered for $p \leq 0.05$.

3. Results and Discussion

3.1. Total Phenolic Content and Antiradical Activity of Red Bean Extract

The extraction of phenolic compounds from red beans was optimized by testing different mixtures of MeOH:H_2O and further determined through the Folin–Ciocalteu assay. The results are shown in Table 1. Based on the results obtained, the mixture 70:30 was optimal to extract phenolic compounds from red beans. The obtained TPC content (2325 ± 98 mg GAE/Kg) was in the same range to that of other studies performed on similar food matrices with values of 1230 mg/Kg [23], 1205 mg/Kg [24], 1690–4850 mg/Kg [25], and 3450 mg/Kg [26]. The TPC and DPPH data of undiluted red bean extracts are shown in Table S1.

Table 1. Total phenolic content (TPC) and antiradical activity (DPPH) of dry red beans using different mixtures of MeOH:H$_2$O. Values are reported as mean $\pm$ SD of independent experiments performed in triplicate.

MeOH:H$_2$O (*v/v*)	TPC (mg GAE/Kg $\pm$ SD)	DPPH (mmol TE/Kg $\pm$ SD)
80:20	1892 $\pm$ 237	42.3 $\pm$ 5.8
70:30	2325 $\pm$ 98	49.2 $\pm$ 4.6
50:50	1348 $\pm$ 24	30.4 $\pm$ 6.1
30:70	1272 $\pm$ 8	29.2 $\pm$ 8.7
20:80	1287 $\pm$ 28	27.5 $\pm$ 7.3

TPC: total phenolic content; GAE: gallic acid equivalents; DPPH: antiradical activity; TE: Trolox equivalents.

The total free radical scavenging activity of the red bean extract was evaluated through the DPPH assay. The results are in agreement with the previous assay being the highest antioxidant activity obtained with the mixture MeOH:H$_2$O (70:30, *v/v*) (Table 1). In addition, it was observed that the larger the phenolic content, the better the free radical scavenging activity. Significant differences in the phenolic content and the antioxidant activity amongst legume extracts were also reported by Zhao et al. [27]. Similarly, the values of antiradical activity obtained from red bean extract (49.2 $\pm$ 4.6 mmol TE/Kg) were comparable to those reported in other studies [28,29].

Besides the antioxidant properties shown by the phenolic compounds, the anti-inflammatory activity by different mechanisms, including modulation of the inflammatory cascade, has also been reported in the literature [30]. García-Lafunete et al. reported that the phenolic rich extracts from beans inhibited the expression of IL-1β, IL-6, and TNF-α genes of stimulated macrophages RAW 246.7, with colored beans showing more activity than white beans [31].

3.2. Identification and Quantification of Active Compounds in Red Bean Extract

A total of 30 polyphenols were investigated in the red beans by using UHPLC-Q-Orbitrap HRMS analysis. The chromatographic and spectrometric parameters are shown in Table 2. The results showed a good chromatographic shape and separation of all studied compounds through the UHPLC gradient system employed within a 13 min run. Four different structural isomers, namely genistein and apigenin (*m/z* 269.0455), and vitexin and apigenin 7-glucoside (*m/z* 431.0983), were found. Compound identification was conducted by comparing the retention times of the standards with the peaks observed in sample extracts. For quantification purposes, calibration curves at eight concentration levels were built in triplicate. All regression coefficients were greater than 0.990.

The quantification of the main phenolic acids and flavonoids in red beans was performed by using a UHPLC-Q-Orbitrap HRMS method. The results are shown in Table 3. Up to 22 polyphenols were detected. Among the quantified polyphenols, epicatechin was the major flavanol (3.297 $\pm$ 0.119 mg/Kg). The delphinidin, an important anthocyanidin of pigmented beans, was quantified at 3.108 $\pm$ 0.023 mg/Kg. Flavanols and anthocyanidins represented 36% and 24.5% of total polyphenolic compounds in red bean samples, respectively. The other important polyphenols quantified in samples were *p*-coumaric acid, isoquercetin, and kaempferol 3-O-glucoside. The polyphenols content of undiluted red bean extracts are shown in Table S2.

Table 2. Chromatographic and spectrometric parameters, including retention time, adduct ion, theoretical and measured mass (m/z), accuracy, and sensibility, for the investigated bioactive compounds ($n = 30$).

Compound	RT (min)	Chemical Formula	Adduct Ion	Theoretical Mass (m/z)	Measured Mass (m/z)	Product Ion	Mass Accuracy (Δ ppm)	LOD (mg/Kg)	LOQ (mg/Kg)
Protocatechuic acid	1.60	$C_7H_6O_4$	$[M-H]^-$	153.01930	153.01857	109.02840	−4.771	0.026	0.078
Cyanidin 3,5-diglucoside	3.03	$C_{27}H_{31}O_{16}$	$[M+H]^+$	611.16066	611.16022	449.19708–287.06469	−0.719	0.026	0.078
Epicatechin	3.08	$C_{15}H_{14}O_7$	$[M-H]^-$	289.07176	289.07202	221.94647–203.09201–161.04478	0.890	0.013	0.039
Chlorogenic acid	3.20	$C_{16}H_{18}O_9$	$[M-H]^-$	353.08780	353.08798	191.05594–84.98998	0.509	0.013	0.039
Cyanidin 3-galactoside	3.23	$C_{21}H_{21}O_{11}$	$[M+H]^+$	449.10784	449.10654	287.05576	−2.894	0.026	0.078
Caffeic acid	3.25	$C_9H_8O_4$	$[M-H]^-$	179.03498	179.03455	134.99960	−2.401	0.013	0.039
Catechin	3.27	$C_{15}H_{14}O_6$	$[M-H]^-$	289.07175	289.07205	247.02241–205.10712–151.03923–125.02335	1.037	0.026	0.078
p-cumaric acid	3.39	$C_9H_8O_3$	$[M-H]^-$	163.04001	163.03937	119.04917	−3.925	0.026	0.078
Apigenin 7-glucoside	3.45	$C_{21}H_{20}O_{10}$	$[M-H]^-$	431.09837	431.09875	341.10919–283.26419	0.881	0.013	0.039
Genistein	3.47	$C_{15}H_{10}O_5$	$[M-H]^-$	269.04554	269.04562	241.14435–213.14908–151.03935	0.297	0.013	0.039
Delphinidin	3.48	$C_{15}H_{11}O_7$	$[M-H]^+$	303.04992	303.04993	257.12119–137.05981	0.033	0.026	0.078
Naringin	3.54	$C_{27}H_{32}O_{14}$	$[M-H]^-$	579.17193	579.17185	271.15524–227.12846–161.04475	−0.138	0.013	0.039
Cyanidin	3.57	$C_{15}H_{11}O_6$	$[M+H]^+$	287.05501	287.05472	207.05879–147.07649	−0.611	0.013	0.039
Rosmarinic acid	3.58	$C_{18}H_{16}O_8$	$[M-H]^-$	359.07724	359.07742	179.05537	0.501	0.013	0.039
Myricitrin	3.59	$C_{21}H_{20}O_{12}$	$[M-H]^-$	463.08820	463.08701	316.02126–178.97646	−2.57	0.013	0.039
Diosmin	3.64	$C_{28}H_{32}O_{15}$	$[M-H]^-$	607.16684	607.16534	300.99796–284.03838	−2.471	0.013	0.039
Isoquercetin	3.65	$C_{21}H_{20}O_{12}$	$[M-H]^-$	463.08820	463.08853	431.09848–187.09698–174.95542	0.712	0.013	0.039
Rutin	3.65	$C_{27}H_{30}O_{16}$	$[M-H]^-$	609.14611	609.14673	300.99911–271.05026–255.12390	1.017	0.013	0.039
Kaempferol 3-glucoside	3.66	$C_{21}H_{20}O_{11}$	$[M-H]^-$	447.09195	447.09329	284.03079–255.02881–227.07033	3.000	0.013	0.039
Vitexin	3.67	$C_{21}H_{20}O_{10}$	$[M-H]^-$	431.09837	431.09711	341.10803–311.05457–269.13815	−2.921	0.013	0.039
Ellagic acid	3.67	$C_{14}H_6O_8$	$[M-H]^-$	300.99899	300.99911	245.91669–229.93712–185.01208–117.00336	0.398	0.013	0.039
Luteolin 7-glucoside	3.68	$C_{21}H_{20}O_{11}$	$[M-H]^-$	447.09328	447.09381	285.04028	1.185	0.013	0.039
Myricetin	3.73	$C_{15}H_{10}O_8$	$[M-H]^-$	317.03029	317.02924	178.87917–151.00217–137.02290	−3.310	0.013	0.039
Daidzein	3.75	$C_{15}H_9O_4$	$[M-H]^-$	253.05063	253.04977	209.96429–225.00984	−3.398	0.013	0.039
Quercetin	3.86	$C_{15}H_{10}O_7$	$[M-H]^-$	301.03538	301.03508	174.95551	−0.996	0.013	0.039
Delphinidin 3,5-diglucoside	3.87	$C_{27}H_{31}O_{17}$	$[M+H]^+$	628.16340	628.16385	465.10339–303.04987	0.716	0.026	0.078
Naringenin	3.91	$C_{15}H_{12}O_5$	$[M-H]^-$	271.06120	271.06110	235.92595–151.03917	−0.368	0.013	0.039

Table 2. *Cont.*

Compound	RT (min)	Chemical Formula	Adduct Ion	Theoretical Mass (*m/z*)	Measured Mass (*m/z*)	Product Ion	Mass Accuracy (Δ ppm)	LOD (mg/Kg)	LOQ (mg/Kg)
Luteolin	3.94	$C_{15}H_{10}O_6$	$[M - H]^-$	285.04046	285.04086	174.95486–89.02095	1.401	0.013	0.039
Kaempferol	4.00	$C_{15}H_{10}O_6$	$[M - H]^-$	285.04046	285.04086	93.00679	1.403	0.013	0.039
Apigenin	4.08	$C_{15}H_{10}O_5$	$[M - H]^-$	269.04554	269.04541	225.06136–117.01828	−0.483	0.013	0.039

RT: retention time; LOD: limit of detection; LOQ: limit of quantification.

Table 3. Polyphenols content in dry red beans. Results are expressed as mean ± SD from three independent determinations.

Compound	Content (mg/Kg) ± SD
Apigenin 7-O-glucoside	<LOQ
Catechin	<LOQ
Chlorogenic acid	0.045 ± 0.002
Cyanidin	0.677 ± 0.046
Cyanidin 3-glucoside	<LOQ
Cyanidin 3,5-diglucoside	0.171 ± 0.015
Daidzein	<LOQ
Delphinidin	3.108 ± 0.023
Delphinidin 3,5-diglucoside	0.210 ± 0.020
Ellagic acid	0.045 ± 0.001
Epicatechin	3.297 ± 0.119
Genistin	<LOQ
Isoquercetin	1.039 ± 0.119
kaempferol 3-O-glucoside	0.803 ± 0.026
Luteolin	0.021 ± 0.001
Naringenin	<LOQ
Naringin	<LOQ
p-coumaric acid	1.929 ± 0.106
Protocatechiuc acid	0.532 ± 0.016
Quercetin	0.292 ± 0.026
Rosmarinic acid	<LOQ
Rutin	0.479 ± 0.037

3.3. Effects of Red Bean Extract on the Cell Viability by Individual Exposure

The HepG2 cell viability evaluated by the MTT assay after 24 h of exposure with red bean extract dilutions, from 1 to 1:32, is shown in Figure 1. The results clearly indicated that the viability of HepG2 cells was affected by the more concentrated red bean extracts. In particular, the undiluted (1) and diluted (1:2 and 1:4) red bean extracts significantly decreased HepG2 cell viability from 83% to 23%. Nonetheless, the red bean extract diluted 1:8 significantly increased the cell viability (20%) compared to the control.

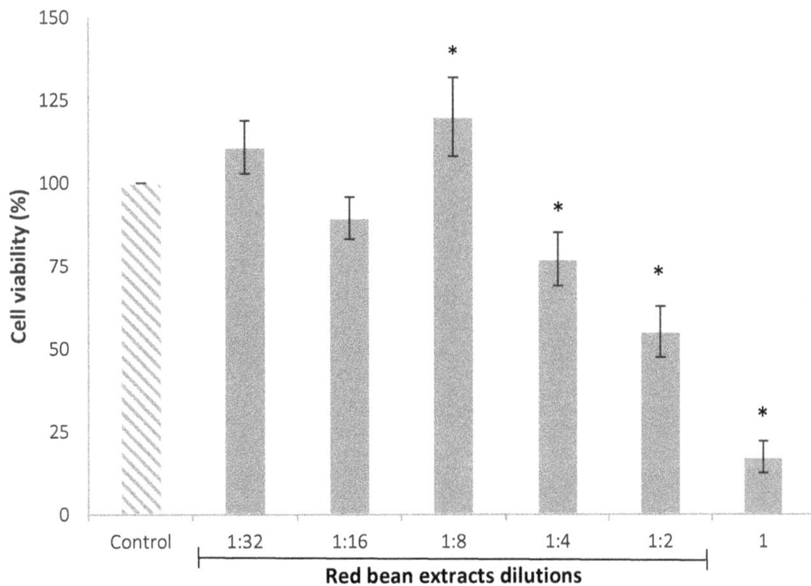

Figure 1. HepG2 cell viability after 24 h of treatment with red bean extracts (undiluted extract (1) and dilutions from 1:2 to 1:32). All values are expressed as mean ± SEM of 3 replicates. (*) $p \leq 0.05$ indicates significant differences compared to control.

The reduction in cell viability caused by the more concentrated or undiluted extracts has also been observed in other studies. Recently, Ziemlewska et al. [32] indicated that high concentrations of the bioactive compounds contained in red fruits inhibited the cell cycle in the G2/M phase and caused cell death, exerting a negative impact on cell viability. These authors demonstrated that high concentrations of phenolic extracts from berries induced apoptosis due to the activation of the caspases.

3.4. Effects of Simultaneous Treatment in HepG2 Cell Viability Exposed to T-2 and Red Bean Extract

The effect in HepG2 cells simultaneously exposed to T-2 (7.5, 15 and 30 nM) and 1:8 diluted red bean extract is described in Figure 2. The HepG2 cell viability was affected in a concentration-dependent manner. Cytotoxic effects were observed in the cells exposed to 30 nM T-2 alone and in combination with the extract, with a significant reduction in cell viability compared to the control cells by 32% and 56%, respectively. This could be due to the fact that the diluted red bean extract did not completely prevent T-2 cytotoxicity. The diluted red bean extract slightly lowered HepG2 cell viability, which was a statistically significant decrease only in cells exposed to the highest T-2 concentration; whereas it did not completely prevent T-2 cytotoxicity.

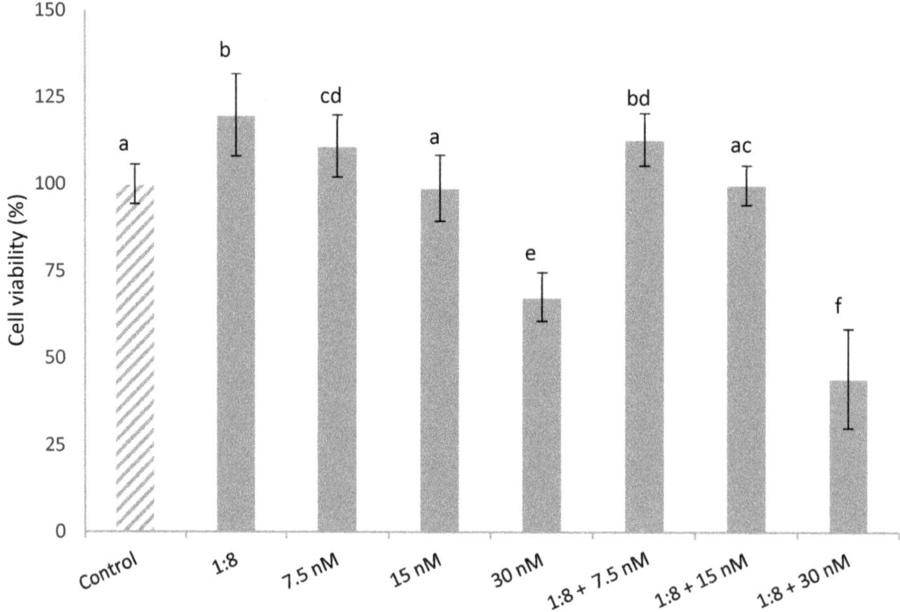

Figure 2. HepG2 cell viability (%) during 24 h of exposure to T-2 (7.5, 15 and 30 nM) and 1:8 diluted red bean extract alone and in combination. All values are expressed as mean $\pm$ SEM of 3 replicates. Values in the same figure with different superscript letters are significantly different ($p \leq 0.05$). 1:8 = 1:8 diluted red bean extract.

On the other hand, cells exposed to the lowest T-2 concentration tested (7.5 nM) showed a significant increase in cell viability (11%) compared to the control; cell viability was slightly higher (13%) in cells simultaneously treated with T-2 and the diluted red bean extract.

3.5. Effects of Pre-Treatment in HepG2 Cell Viability Exposed to T-2 and Red Bean Extract

The effects of cell pre-treatment with red bean extract for 1 h and 24 h before the T-2 addition are shown in Figure 3. The results indicated that the pre-treatment with red bean extract was not able to protect or ameliorate the cytotoxic T-2 effect in HepG2 cells in all T-2 concentrations tested. This effect was increased at 24 h of exposure. Similar results were reported by Kössler et al. [33], who showed that curcumin (phenolic compound) reduced cell viability and induced apoptosis in human embryonic kidney cells (HEK293) in a dose-dependent manner at 22 h of exposure. There are no other data from in vitro studies by other authors with bean extracts. However, Vila-Donat et al. [34] reported synergistic cytoprotective effects against cytotoxicity induced by alternariol on Caco-2 cells exposed to an extract obtained from other types of legumes, such as lentils.

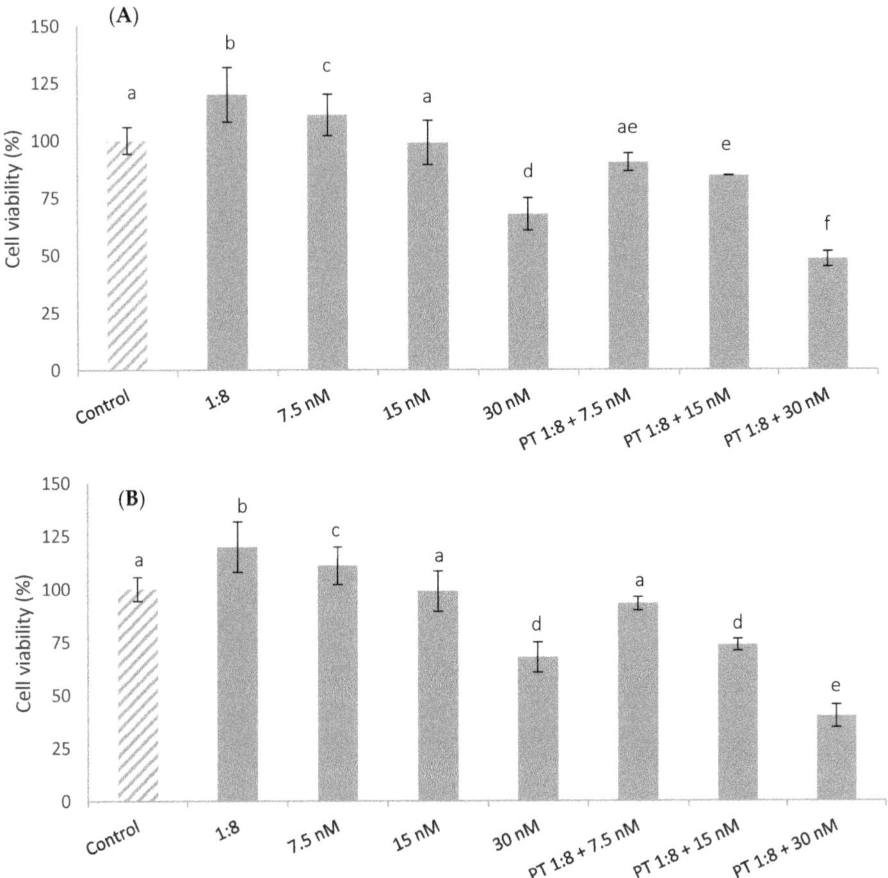

Figure 3. HepG2 cell viability (%) of pre-treated cells (PT) with 1:8 red bean extract for 1 h (**A**) and 24 h (**B**) and afterward exposed to T-2 (7.5, 15, and 30 nM) during 24 h. All values are expressed as mean ± SEM of 3 replicates. Values in the same figure with different superscript letters are significantly different ($p \leq 0.05$).

3.6. Reactive Oxygen Species

The intracellular accumulation of ROS in HepG2 cells exposed to T-2 toxin was analyzed using the dichlorofluorescein assay (DCFH-DA). The cells were exposed to T-2 (7.5, 15, and 30 nM) and 1:8 diluted red bean extract for different exposure times (0, 5, 15, 30, 45, 60, 90, and 120 min), and the results are expressed as an increase in fluorescence (%) with respect to the control (Figure 4). The simultaneous exposure to T-2 at the lowest concentration tested (7.5 nM) and the red bean extract showed a significant reduction ($p \leq 0.05$) in ROS production after 120 min of exposure with respect to control cells (Figure 4A); whereas there were no significant differences in ROS production at 15 nM with respect to control cells (Figure 4B). On the contrary, a significant increase ($p \leq 0.05$) in ROS production with respect to controls was observed in the case of the simultaneous exposure to T-2 at 30 nM alone and in combination with the red bean extract at 60 min of exposure (Figure 4C).

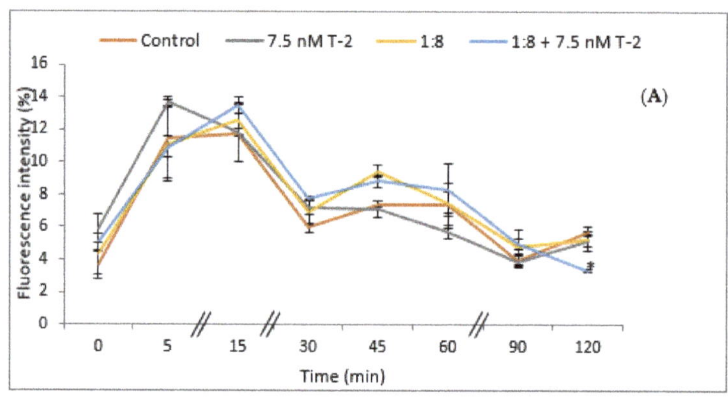

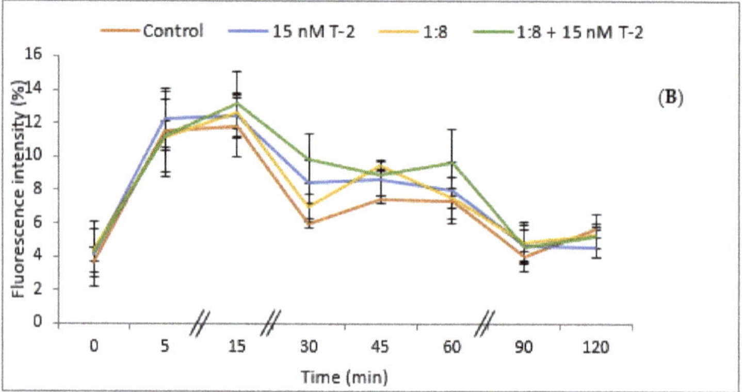

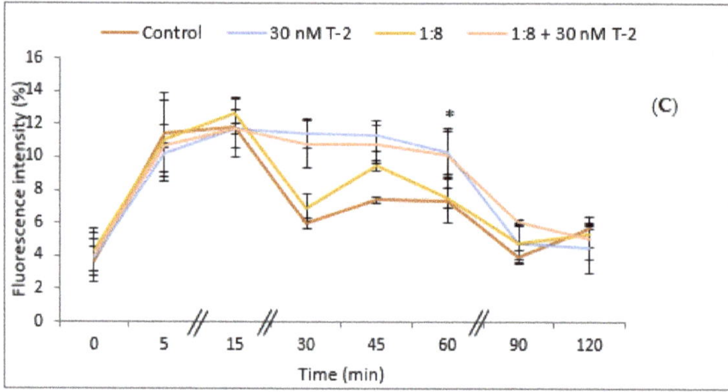

Figure 4. Intracellular ROS production in HepG2 cells exposed to: (**A**) 7.5 nM T-2; (**B**) 15 nM T-2; (**C**) 30 nM T-2; and 1:8 diluted red bean extract, alone or in combination, after 0, 5, 15, 30, 45, 60, 90, and 120 min. All values are expressed as mean ± SEM of 4 replicates. (*) $p \leq 0.05$ represents significant difference as compared to control values.

Data reported by Lee et al. [34] demonstrated that the treatment of HepG2 cells with extracts obtained from different types of beans (black, red, and green) against the Tert Butyl Hydroperoxide (TBHP) agent significantly decreased the production of ROS. On the other hand, Yang et al. [3] showed that the level of intracellular ROS production in HCT116 cells decreased significantly in cells treated with phenolic extracts of beans compared to untreated cells.

The abovementioned results can demonstrate the ability of phenolic extracts to modulate oxidative stress. However, further studies with regard to biological activity, including studies into the mechanisms of action and structure-activity relationships, are necessary to fully understand the modes of action of these bioactive compounds and to fully exploit their cytoprotective potential effect, as highlighted in the literature [35].

4. Conclusions

The chemical profile obtained from this study evidenced that red beans are a rich source of bioactive compounds such as flavanols and anthocyanidins, which confers anti-radical activity and human health benefits. On the other hand, red bean extract (diluted 1:8) showed a significant increase in HepG2 cell proliferation after 24 h of exposure. Similar results were observed after HepG2 cell exposure to T-2 toxin at the lowest concentration assayed (7.5 nM) corresponding to its $IC_{50}/8$, which could indicate a probable adaptative response of HepG2 cells. However, higher T-2 concentrations showed cytotoxic effects and ROS production on HepG2 cells. The results from the simultaneous and pre-treatment assays indicated that the 1:8 diluted red bean extract did not prevent T-2 cytotoxicity neither in simultaneous exposures nor with the pre-treatment. Finally, the combination of T-2 at 7.5 nM with the diluted red bean extract showed a decrease in ROS production compared to the control at the longest exposure time tested (120 min). The antioxidant activity or the possible T-2 hormetic effect observed in simultaneous treatment could be responsible for the latter result, suggesting that the red bean extract could modulate oxidative stress on HepG2 cells.

Supplementary Materials: The following supporting information can be downloaded at: https://www.mdpi.com/article/10.3390/foods11071033/s1, Table S1: Total phenolic content (TPC) and antiradical activity (DPPH) of undiluted red bean extracts using different mixtures of MeOH:H_2O. Values are reported as mean $\pm$ SD of independent experiments performed in triplicate; Table S2. Polyphenols content in undiluted red bean extracts. Results are expressed as mean $\pm$ SD from three independent determinations.

Author Contributions: Conceptualization, Y.R.-C., M.-J.R. and A.R.; methodology, C.M.-A., M.T. and L.C.; software, M.T. and L.I.; validation, M.T., L.I. and Y.R.-C.; formal analysis, C.M.-A., M.T., L.C., L.I.; writing—original draft preparation, C.M.-A., L.I., Y.R.-C.; writing—review and editing, M.-J.R., A.R. and Y.R.-C.; supervision, A.R. and M.-J.R.; project administration, M.-J.R.; funding acquisition, M.-J.R. All authors have read and agreed to the published version of the manuscript.

Funding: This research was funded by the Spanish Ministry of Science and Innovation Project (PID2020-115871RB-I00).

Institutional Review Board Statement: Not applicable.

Informed Consent Statement: Not applicable.

Data Availability Statement: The data presented in this study are available on request from the corresponding author. The data are not publicly available to preserve the privacy of the volunteers that participated in the present study.

Conflicts of Interest: The authors declare no conflict of interest.

References

1. Rahate, K.; Madhumita, M.; Prabhakar, P. Nutritional composition, anti-nutritional factors, pretreatments-cum-processing impact and food formulation potential of faba bean (*Vicia faba* L.): A comprehensive review. *LWT-Food Sci. Technol.* **2021**, *138*, 110796. [CrossRef]
2. Angiolillo, L.; Del Nobile, M.A.; Conte, A. The extraction of bioactive compounds from food residues using microwaves. *Curr. Opin. Food Sci.* **2015**, *5*, 93–98. [CrossRef]
3. Yang, Q.; Gan, R.; Ge, Y.; Zhang, D.; Corke, H. Polyphenols in Common Beans (*Phaseolus vulgaris* L.): Chemistry, Analysis, and Factors Affecting Composition. *Compr. Rev. Food Sci. Food Saf.* **2018**, *17*, 1518–1539. [CrossRef] [PubMed]

4. Gan, R.; Deng, Z.; Yan, A.; Shah, N.; Lui, W.; Chan, C.; Corke, H. Pigmented edible bean coats as natural sources of polyphenols with antioxidant and antibacterial effects. *LWT-Food Sci. Technol.* **2016**, *73*, 168–177. [CrossRef]
5. Kimothi, S.; Dhaliwal, Y. Nutritional and Health Promoting Attribute of Kidney Beans (*Phaseolus vulgaris* L.): A Review. *Int. J. Curr. Microbiol. Appl. Sci.* **2020**, *9*, 1201–1209. [CrossRef]
6. Taroncher, M.; Vila-Donat, P.; Tolosa, J.; Ruiz, M.J.; Rodríguez-Carrasco, Y. Biological activity and toxicity of plant nutraceuticals: An overview. *Curr. Opin. Food Sci.* **2021**, *42*, 113–118. [CrossRef]
7. Madhujith, T.; Naczk, M.; Shahidi, F. Antioxidant activity of common beans (*Phaseolus vulgaris* L.). *J. Food Lipids* **2004**, *11*, 220–233. [CrossRef]
8. Khan, N.; Monagas, M.; Llorach, R.; Urpi-Sarda, M.; Rabassa, M.; Estruch, R.; Andrés-Lacueva, C. Targeted and metabolomic study of biomarkers of cocoa powder consumption effects on inflammatory biomarkers in patients at high risk of cardiovascular disease. *Agro Food Ind. Hi Tech.* **2010**, *21*, 51–54.
9. Terra, X.; Montagut, G.; Bustos, M.; Llopiz, N.; Ardèvol, A.; Bladé, C.; Fernández-Larrea, J.; Pujadas, G.; Salvadó, J.; Arola, L.; et al. Grape-seed procyanidins prevent low-grade inflammation by modulating cytokine expression in rats fed a high-fat diet. *J. Nutr. Biochem.* **2009**, *20*, 210–218. [CrossRef] [PubMed]
10. Al-Hanbali, M.; Ali, D.; Bustami, M.; Abdel-Malek, S.; Al-Hanbali, R.; Alhussainy, T.; Qadan, F.; Matalka, K.Z. Epicatechin suppresses IL-6, IL-8 and enhances IL-10 production with NF-kappaB nuclear translocation in whole blood stimulated system. *Neuro Endocrinol. Lett.* **2009**, *30*, 131–138. [PubMed]
11. RASFF (Rapid Alert for Food and Feed). RASFF Annual Report 2020. Available online: https://ec.europa.eu/food/safety/rasff-food-and-feed-safety-alerts/reports-and-publications_es (accessed on 2 February 2022).
12. Narváez, A.; Castaldo, L.; Izzo, L.; Pallarés, N.; Rodríguez-Carrasco, Y.; Ritieni, A. Deoxynivalenol contamination in cereal-based foodstuffs from Spain: Systematic review and meta-analysis approach for exposure assessment. *Food Control* **2021**, *132*, 108521. [CrossRef]
13. EFSA (European Food Safety Authority). Scientific report on Human and animals dietary exposure to T-2 and HT-2 toxin. *EFSA J.* **2017**, *15*, 4972.
14. Dai, C.; Xiao, X.; Sun, F.; Zhang, Y.; Hoyer, D.; Shen, J.; Tang, S.; Velkov, T. T-2 toxin neurotoxicity: Role of oxidative stress and mitochondrial dysfunction. *Arch. Toxicol.* **2019**, *93*, 3041–3056. [CrossRef] [PubMed]
15. You, L.; Zhao, Y.; Kuca, K.; Wang, X.; Oleksak, P.; Chrienova, Z.; Nepovimova, E.; Jaćević, V.; Wu, Q.; Wu, W. Hypoxia, oxidative stress, and immune evasion: A trinity of the trichothecenes T-2 toxin and deoxynivalenol (DON). *Arch. Toxicol.* **2021**, *95*, 1899–1915. [CrossRef]
16. Taroncher, M.; Rodríguez-Carrasco, Y.; Ruiz, M. Interactions between T-2 toxin and its metabolites in HepG2 cells and in silico approach. *Food Chem. Toxicol.* **2021**, *148*, 111942. [CrossRef] [PubMed]
17. Vila-Donat, P.; Fernández-Blanco, C.; Sagratini, G.; Font, G.; Ruiz, M.J. Effects of soyasaponin I and soyasaponins-rich extract on the alternariol-induced cytotoxicity on Caco-2 cells. *Food Chem. Toxicol.* **2015**, *77*, 44–49. [CrossRef]
18. Izzo, L.; Rodríguez-Carrasco, Y.; Pacifico, S.; Castaldo, L.; Narváez, A.; Ritieni, A. Colon Bioaccessibility under In Vitro Gastrointestinal Digestion of a Red Cabbage Extract Chemically Profiled through UHPLC-Q-Orbitrap HRMS. *Antioxidants* **2020**, *9*, 955. [CrossRef]
19. Castaldo, L.; Izzo, L.; Gaspari, A.; Lombardi, S.; Rodríguez-Carrasco, Y.; Narváez, A.; Grosso, M.; Ritieni, A. Chemical Composition of Green Pea (Pisum sativum L.) Pods Extracts and Their Potential Exploitation as Ingredients in Nutraceutical Formulations. *Antioxidants* **2022**, *11*, 105. [CrossRef]
20. Taroncher, M.; Rodríguez-Carrasco, Y.; Ruiz, M. T-2 toxin and its metabolites: Characterization, cytotoxic mechanisms and adaptive cellular response in human hepatocarcinoma (HepG2) cells. *Food Chem. Toxicol.* **2021**, *145*, 111654. [CrossRef]
21. Ruiz, M.J.; Festila, L.E.; Fernández, M. Comparison of basal cytotoxicity of seven carbamates in CHO-K1 cells. *Toxicol. Environ. Chem.* **2006**, *88*, 345–354. [CrossRef]
22. Ruiz-Leal, M.; George, S. An in vitro procedure for evaluation of early stage oxidative stress in an established fish cell line applied to investigation of PHAH and pesticide toxicity. *Mar. Environ. Res.* **2004**, *58*, 631–635. [CrossRef] [PubMed]
23. Soriano Sancho, R.; Pavan, V.; Pastore, G. Effect of in vitro digestion on bioactive compounds and antioxidant activity of common bean seed coats. *Food Res. Int.* **2015**, *76*, 74–78. [CrossRef]
24. Ombra, M.N.; d'Acierno, A.; Nazzaro, F.; Riccardi, R.; Spigno, P.; Zaccardelli, M.; Pane, C.; Maione, M.; Fratanni, F. Phenolic Composition and Antioxidant and Antiproliferative Activities of the Extracts of Twelve Common Bean (*Phaseolus vulgaris* L.) Endemic Ecotypes of Southern Italy before and after Cooking. *Oxid. Med. Cell. Longev.* **2016**, *12*, 1398298. [CrossRef] [PubMed]
25. Orak, H.H.; Karamać, M.; Amarowicz, R. Antioxidant activity of phenolic compounds of red bean (*Phaseolus vulgaris* L). *Oxid. Commun.* **2015**, *38*, 67–76.
26. Ross, K.; Beta, T.; Arntfield, S. A comparative study on the phenolic acids identified and quantified in dry beans using HPLC as affected by different extraction and hydrolysis methods. *Food Chem.* **2009**, *113*, 336–344. [CrossRef]
27. Zhao, Y.; Du, S.; Wang, H.; Cai, M. In vitro antioxidant activity of extracts from common legumes. *Food Chem.* **2014**, *152*, 462–466. [CrossRef]
28. Korus, J.; Gumul, D.; Fołta, M.; Bartoń, H. Antioxidant and antiradical activity of raw and extruded common beans. *Electron. J. Pol. Agric. Univ.* **2007**, *10*, 6.

29. Doğan Cömert, E.; Ataç Mogol, B.; Gökmen, V. Relationship between color and antioxidant capacity of fruits and vegetables. *Curr. Res. Food Sci.* **2020**, *2*, 1–10. [CrossRef]
30. García-Lafuente, A.; Guillamón, E.; Villares, A.; Rostagno, M.; Martínez, J.A. Flavonoids as anti-inflammatory agents: Implications in cancer and cardiovascular disease. *Inflamm. Res.* **2009**, *58*, 537–552. [CrossRef]
31. García-Lafuente, A.; Moro, C.; Manchón, N.; Gonzalo-Ruiz, A.; Villares, A.; Guillamón, E.; Rostagno, M.; Mateo-Vivaracho, L. In vitro anti-inflammatory activity of phenolic rich extracts from white and red common beans. *Food Chem.* **2014**, *161*, 216–223. [CrossRef]
32. Ziemlewska, A.; Zagórska-Dziok, M.; Nizioł-Łukaszewska, Z. Assessment of cytotoxicity and antioxidant properties of berry leaves as by-products with potential application in cosmetic and pharmaceutical products. *Sci. Rep.* **2021**, *11*, 3240. [CrossRef] [PubMed]
33. Kössler, S.; Nofziger, C.; Jakab, M.; Dossena, S.; Paulmichl, M. Curcumin affects cell survival and cell volume regulation in human renal and intestinal cells. *Toxicology* **2012**, *292*, 123–135. [CrossRef] [PubMed]
34. Lee, J.H.; Ham, H.; Kim, M.Y.; Ko, J.Y.; Sim, E.; Kim, H.; Lee, C.K.; Jeon, Y.H.; Jeong, H.S.; Woo, K.S. Phenolic compounds and antioxidant activity of adzuki bean cultivars. *Legume Res.* **2018**, *41*, 681–688. [CrossRef]
35. Zhu, F.; Du, B.; Xu, B. Anti-inflammatory effects of phytochemicals from fruits, vegetables, and food legumes: A review. *Crit Rev. Food Sci. Nutr.* **2018**, *58*, 1260–1270. [CrossRef] [PubMed]

Article

Metabolomic Analysis Reveals Nutritional Diversity among Three Staple Crops and Three Fruits

Yunxia Shi [1], Yanxiu Guo [1], Yuhui Wang [1], Mingyang Li [1], Kang Li [1,2], Xianqing Liu [1,2], Chuanying Fang [1] and Jie Luo [1,2,*]

1   School of Tropical Crops, Hainan University, Haikou 570288, China; yunxiashi225@163.com (Y.S.); gyx13563502250@163.com (Y.G.); wyh0140@163.com (Y.W.); lygaugau@163.com (M.L.); kang_li@hainanu.edu.cn (K.L.); liuxq@hainanu.edu.cn (X.L.); cyfang@hainanu.edu.cn (C.F.)
2   Hainan Yazhou Bay Seed Laboratory, Sanya Nanfan Research Institute of Hainan University, Sanya 572025, China
*   Correspondence: jie.luo@hainanu.edu.cn

Abstract: More than 2 billion people worldwide are under threat of nutritional deficiency. Thus, an in-depth comprehension of the nutritional composition of staple crops and popular fruits is essential for health. Herein, we performed LC-MS-based non-targeted and targeted metabolome analyses with crops (including wheat, rice, and corn) and fruits (including grape, banana, and mango). We detected a total of 2631 compounds by using non-targeted strategy and identified more than 260 nutrients. Our work discovered species-dependent accumulation of common present nutrients in crops and fruits. Although rice and wheat lack vitamins and amino acids, sweet corn was rich in most amino acids and vitamins. Among the three fruits, mango had more vitamins and amino acids than grape and banana. Grape and banana provided sufficient 5-methyltetrahydrofolate and vitamin B6, respectively. Moreover, rice and grape had a high content of flavonoids. In addition, the three crops contained more lipids than fruits. Furthermore, we also identified species-specific metabolites. The crops yielded 11 specific metabolites, including flavonoids, lipids, and others. Meanwhile, most fruit-specific nutrients were flavonoids. Our work discovered the complementary pattern of essential nutrients in crops and fruits, which provides metabolomic evidence for a healthy diet.

Keywords: crops; fruits; nutrition; metabolite analyses; metabolome

Citation: Shi, Y.; Guo, Y.; Wang, Y.; Li, M.; Li, K.; Liu, X.; Fang, C.; Luo, J. Metabolomic Analysis Reveals Nutritional Diversity among Three Staple Crops and Three Fruits. *Foods* **2022**, *11*, 550. https://doi.org/10.3390/foods11040550

Academic Editors: Yelko Rodríguez-Carrasco and Bojan Šarkanj

Received: 27 January 2022
Accepted: 13 February 2022
Published: 15 February 2022

Publisher's Note: MDPI stays neutral with regard to jurisdictional claims in published maps and institutional affiliations.

1. Introduction

A healthy diet containing enough nutrients is vital to health. The lack of macro- and micro-nutrients in diets threatens health [1], especially in underdeveloped countries where people cannot afford varied diets [2,3]. Rice, wheat, and corn are globally major staple crops. Although they provide 60% of the world's calorie intake [4], the micro-nutrients available from staple foods are limited. For instance, corn lacks tryptophan and lysine, two essential amino acids for humans [5]. The processing of rice and wheat removes bran and embryo, which leads to the loss of most nutrients [6,7].

The benefits of a plant-based diet on health are mainly from the composition of phytochemicals [8,9]. More fruit uptake can reduce the risk of some diseases for the enrichment of vitamins and other micronutrients [10,11]. Mango, banana, and grape are globally the most widely cultivated fruits [12,13]. Mango is rich in phytosterols, carotenoids, and vitamins [14,15]. Phytochemicals in mango protect humans from diabetes, obesity, and cancer [16]. Banana contains high levels of provitamin A carotenoids (pVACs), which alleviate vitamin A deficiency and reduce the risk of cancers and heart diseases [17]. As a popular fruit, grape is famous for highly bioactive phenols, such as anthocyanins, flavanols, flavonols, and resveratrol. These compounds are pivotal in antioxidant, heart protection, anti-cancer, and anti-aging activity [18,19]. Therefore, through food selection and a healthy diet, people can get enough nutrition and reduce the occurrence of nutritional deficiency. However, metabolic cues for food selection are limited.

Metabolomic studies provide novel insights into the composition of phytochemicals. Metabolic diversity among staple food crops has been well documented. For instance, metabolome signatures have varied across developmental stages and genotypes in rice, maize, and wheat [20–22]. Multiple-omics studies have also provided deep insights into genetic bases of metabolic diversity in staple crops [23–25]. Recently, metabolic features of fruits have attracted more and more interest. For example, citrus fruits are primary fruit sources of poly methoxy flavonoids (PMFs), and the determination of major PMFs in fruits or leaves of 116 citrus accessions revealed significant species-specific and spatiotemporal characteristics. All reticulated citrus and its natural or artificial hybrids had detectable PMFs, especially in fruits of wild or early cultivated citrus in early fruit development [26]. In peach, metabolomic analysis was used to construct the metabolic network of peach mesocarp throughout development. In the early developmental stages of peach, protein abundance was significantly reduced, while bioactive polyphenols and amino acids piled up [27]. However, the diversity of nutritional metabolites between crops and fruits is largely unknown.

Here, we performed non-targeted and targeted metabolic profiling to dissect the metabolic diversity by using rice, wheat, corn, banana, grape, and mango. We detected over 3000 metabolites. Those include amino acids, vitamins, flavonoids, lipids, and other nutrients. In addition, we found significant differences in nutrient construction between crops and fruits. Although crops are rich in lipids and sweet corn is rich in vitamins and amino acids, rice and wheat lack in vitamins and amino acids. Grape is rich in flavonols and anthocyanin, but it lacks amino acids. Mango is rich in vitamins, especially vitamin C, and most of amino acids. Banana is rich in vitamin B6 but short in flavonoids. Based on the rich metabolic diversity between crops and fruits, our work deepens our comprehension of the nutrient structure of a diet.

2. Materials and Methods

2.1. Plant Materials

To study the difference in metabolites between crops and fruits, we selected three important crops and three popular fruits for research. Crops included wheat, corn, and rice. Fruits included grape, mango, and banana. Rice (ZH11) came from the breeding base of Hainan University. Corn (sweet corn) came from Ding'an breeding base of Hainan University. Wheat (Chinese Spring) came from common wheat varieties grown in the experimental station of the Institute of genetics and developmental biology, Chinese Academy of Sciences (IGDB, CAS) from 2018 to 2019. Grapes (Pinor Vermei, one of the most popular grape varieties in the world), mangoes (Alphonso, a traditional Indian cultivar) and bananas (Cavendish) were selected from the field germplasm community of Guangzhou banana garden.

2.2. Chemical Reagents

Chromatographic-grade acetonitrile, acetic acid, and methanol were purchased from Merck (Darmstadt, Germany). The Milli-Q water was purified using a Millipore purification system (Millipore Corporation, Burlington, MA, USA). All standards used in the test were stored in −80 °C refrigerator in the dark.

2.3. Metabolite Sample Preparation

Fresh fruit or dry grain of crop were collected into 50 mL centrifuge tubes and quickly frozen in liquid nitrogen and freeze dried. Three biological replicates were collected for each species. The samples were ground into powder using a grinder machine (MM400, Retsch) with steel balls at 28 Hz for 56 s or more. Then, 0.05–0.1 g of sample powder was suspended with 70% methanol water solution in the ratio of 1:10,000. Next, the samples were extracted by ultrasonic wave for 10 min at 50 Hz for a total of three times [22,28]. At the end of each time, vortex vibration and mixing were required.

2.4. Metabolomic Detection

Non-targeted metabolic profiling analyses were performed with Q Exactive Focus Orbitrap LC-MS/MS (Thermo Scientific, Waltham, MA, USA). Scanning mass ranged from m/z 100–1000 with an accumulation time of 0.10 s. The scanning mode was full MS/ddMS2. The recorded data were processed with compound discoverer (CD) 3.1 software to obtain the mass to charge ratio, retention time, MS/MS2 information of all detected substances. Then, the detected signals were automatically matched through the internally established reference libraries of chemical standard entries of software to predict and identify the metabolite information. The multiple reaction monitoring (MRM) mode with QTRAP 6500$^+$ LC-MS/MS (Shimadzu, Kyoto, Japan) was used for targeted metabolome analyses. The detection window was set to 80 s, and the targeted scanning time was 1.5 s. The original data were processed by Multi Quant 3.0.3 software. The chromatographic column was C18 column (Shim-pack GLSS C18, 1.9UM, 2.1*100, Shimadzu). Mobile phase A and B was 0.04% acetic acid–water solution, and mobile phase B was 0.04% acetic acid–methanol solution. The qualitative and quantitative chromatographic conditions were consistent.

2.5. Statistical Analysis

The relative signal strength of metabolites was divided and normalized according to the internal standard (0.1 mg L^{-1} lidocaine), and log2 was then used to transform the value. We used Student's t-test and fold change of difference to screen for differentially accumulated metabolites (DAMs). Metabolites with $p < 0.05$ and |log2 (fold change) | $\geq$ 1 were considered as DAMs. The differences between the metabolites in six fruits were calculated by nested ANOVA in the R package.

3. Results

3.1. Metabolic Analysis of Crops and Fruits

To dissect the diversity of metabolites between crops and fruits, we selected three staple food crops (rice, corn, and wheat) and three popular fruits (mango, grape, and banana). Through non-targeted metabolome detection, we detected a total of 13,790 metabolic signals in six species (Figure 1a). Among them, 7831 signals were detected in rice, 8325 signals in wheat, 8879 signals in corn, 8074 signals in mango, 8135 signals in grape, and 9139 signals in banana (Table S1). Next, we performed a principal component analysis (PCA) of all samples based on the liquid chromatography-mass spectrometry (LC-MS) data. PCA diagram showed that principal component (PC) 1 and 2 explained 32.17% and 21.9% variability, respectively (Figure 1b). Principal component 1 separated crops and fruits successfully, indicating that the diversity of metabolites between crops and fruits was significant.

To further study the metabolic feature of crops and fruits, we quantified the metabolites by scheduled multiple reaction monitoring (SMRM) and finally detected 2631 metabolites. We detected 2296, 2251, 2432, 2290, 2249, and 2376 metabolites in wheat, rice, corn, mango, grape, and banana, respectively (Table S2). Among them, 1963 metabolites were shared by crops and fruits. Meanwhile, 9, 7, 67, 18, 9, and 44 metabolites only existed in wheat, rice, corn, mango, grape, and banana, respectively (Figure 1c). The distribution of the metabolites of crops and fruits is complementary. Substances with high content in crops are relatively low in fruits, and vice versa (Figure 1d).

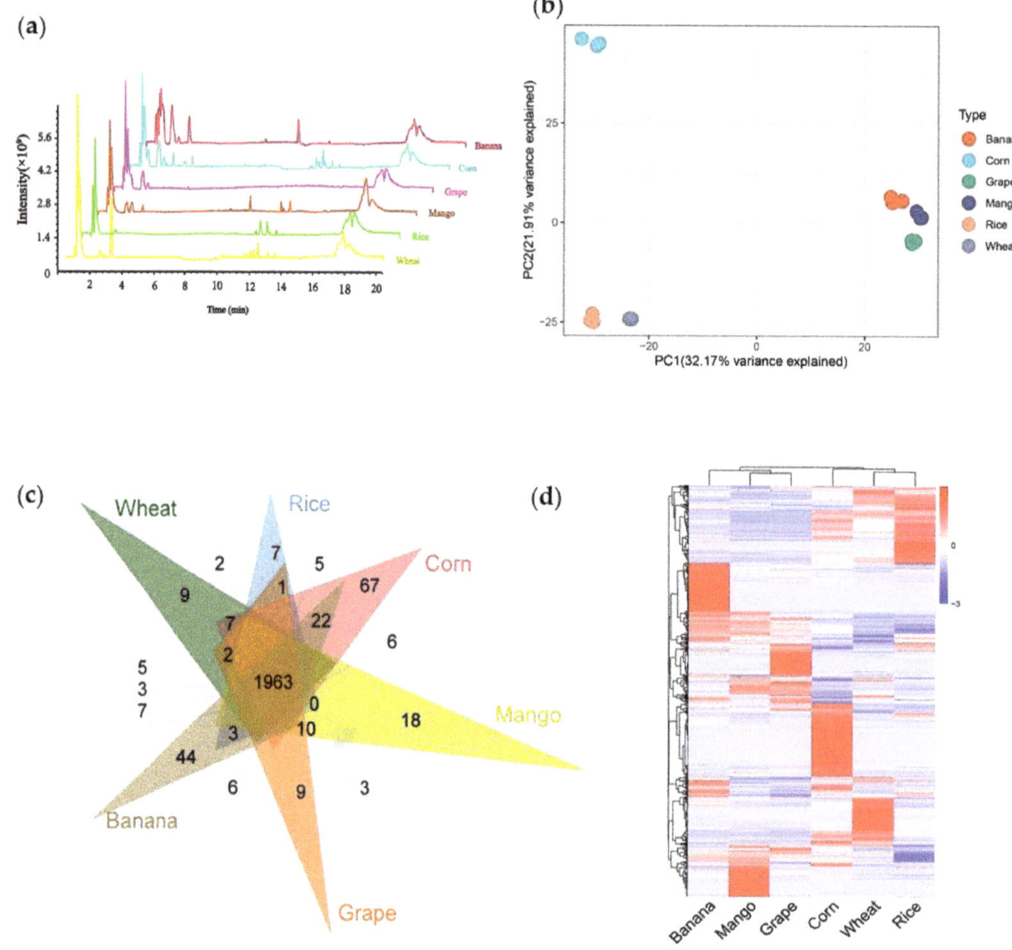

Figure 1. Analysis of metabolic variation in fruits and crops using Q Exactive Focus Orbitrap LC-MS/MS. (**a**) Total ion chromatography of metabolites in fruits and crops. (**b**) Principal component analysis (PCA) of the total ion chromatography of fruits and crops. (**c**) Venn diagram analysis of crops and fruits. (**d**) Heat map analysis of 2631 metabolites detected in crops and fruits.

3.2. Characterization of Metabolic Signals

We characterized species-specific metabolites by the retention time, the relative abundance of fragments, and the mass loss during fragmentation. Then, according to these characteristics, we checked the fragment information in literature and databases, such as mass bank [29] and the Human Metabolome Database (HMDB) [30]. Finally, we annotated some of the metabolites with the help of standards. At the same time, by using Compound Discoverer (CD) 3.1, we matched metabolic structures through the CD database (Figure 2a).

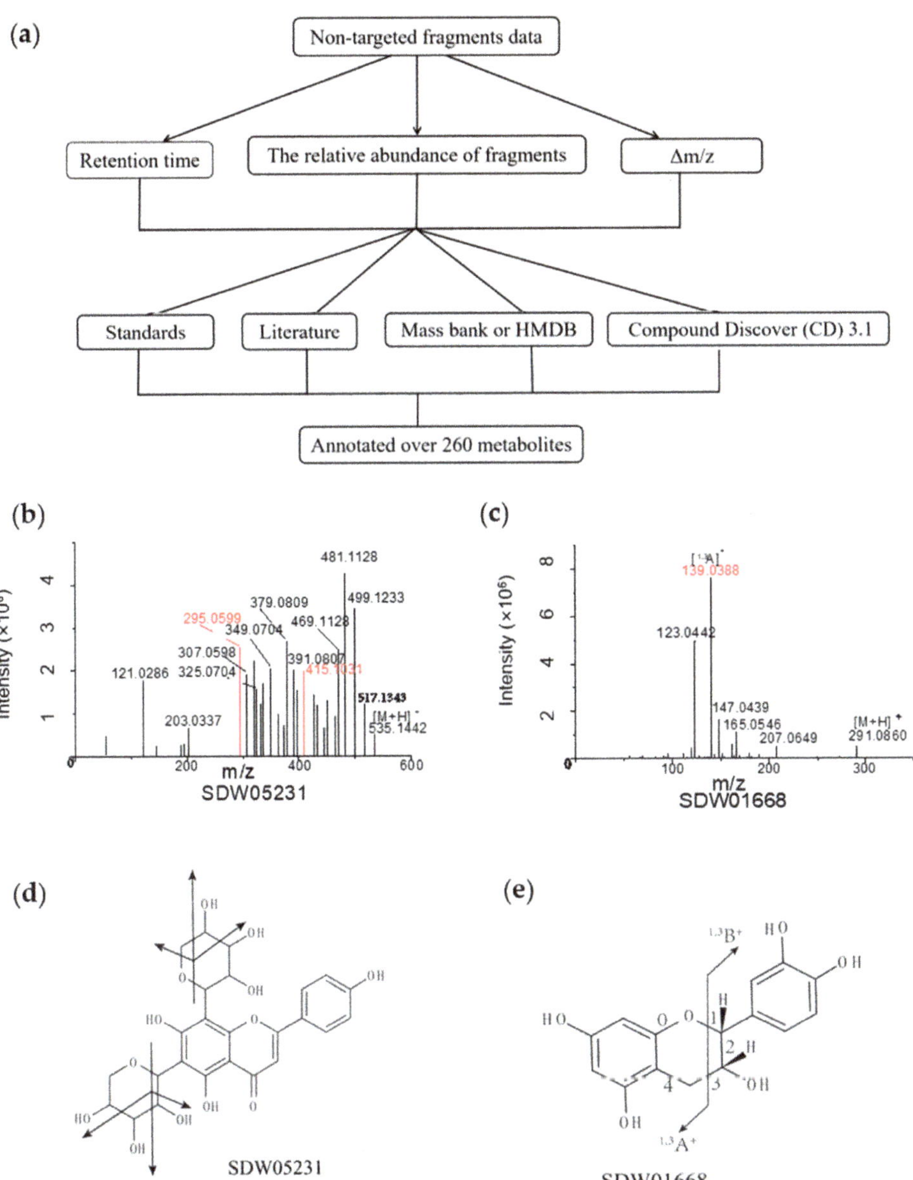

Figure 2. Detection and identification of specific metabolite signs by Q Exactive Focus Orbitrap LC-MS/MS. (**a**) Flowchart for detection and identification of specific metabolites. (**b**) MS/MS spectra of SDW05231 at m/z 535.1442, the metabolite was identified as 5,7-dihydroxy-2-(4-hydroxyphenyl)-6,8-bis(3,4,5-trihydroxyoxan-2-yl)-4H-chromen-4-one. (**c**) MS/MS spectra of SDW01668 at m/z 291.0860, the metabolite was identified as (-)-Epicatechin. (**d**) The molecular structure of the 5,7-dihydroxy-2-(4-hydroxyphenyl)-6,8-bis(3,4,5-trihydroxyoxan-2-yl)-4H-chromen-4-one and its general fragmentation rules. (**e**) The molecular structure of the (-)-Epicatechin and its general fragmentation rules.

The crop-specific SDW05231 (RT 6.26 min) yielded a precursor ion [M+H]$^+$ at *m/z* 535.1442. We observed product ions with uneven abundance in the secondary mass spectrum. The precursor ion lost 18 (H$_2$O) or 30 (CH$_2$O) and produced two signals at *m/z* 415.1031 [M+H-120]$^+$ and *m/z* 295.0599 [M+H-120]$^+$ (red peak spectrum in Figure 2b). Its fragmentation pattern resembled the flavone C-glucoside, which contains two pentose residues. Considering the product ion at *m/z* 295.0599, we inferred that SDW05231 was an apigenin derivative. Therefore, we decoded DWZP05231 as 5,7-dihydroxy-2-(4-hydroxyphenyl)-6,8-bis(3,4,5-trihydroxyoxan-2-yl)-4H-chromen-4-one (Figure 2b,d).

Then, we annotated a fruit-specific SDW01668 as (-)-Epicatechin using the CD database. SDW01668 (RT 4.73 min) produced a precursor ion [M+H]$^+$ at *m/z* 291.0860. The tandem mass spectrum showed a high-intensity fragment [1,3A]$^+$ ion at *m/z* 139.0388. A further loss of 16 Da yielded *m/z* 123.0442 based on [1,3A]$^+$. The secondary fragments' *m/z* and fracture modes were highly similar to (-)-Epicatechin (Figure 2c,e).

We annotated over 260 metabolites (Table S3), 28 and 18 of which were absent in fruits and crops, respectively. Moreover, we also identified species-specific metabolites, such as resveratrol in grapes (Table S4).

3.3. Whole Metabolome Scale Comparative Analysis of Crops and Fruits

To explore the diversity of nutrients between crops and fruits, we performed LC-MS/MS-based targeted metabolome analyses. Firstly, we quantified the metabolites by scheduled multiple reaction monitoring (SMRM). We detected 660 metabolites, including 26 vitamins, 106 amino acids and derivatives, 225 lipids, 108 flavonoids, 73 organic acids, and other metabolites (Table S5). A PCA showed that PC1 and PC2 explained 32.97% and 20.5% of the variability, respectively (Figure 3a). PC1 separated crops and fruits, indicating the significant metabolic difference between them. According to PC2, grape, mango, rice, and wheat clustered together. Meanwhile, banana and corn got close to each other and were far away from the others. The species-dependent accumulation pattern was further visualized by a heatmap based on crop and fruit metabolome data. The six species formed two clusters: crops and fruits (Figure 3b). That is, the metabolic features of crops and fruits differed remarkably.

Then, we analyzed differentially accumulated metabolites (DAMs) between each fruit with three crop species. DAMs between each species pair met the following criterion: the fold change >2, while the *p*-value <0.05. Grapes, for example, accumulated 341~346 DAMs compared with crops. A total of 346 DAMs existed between rice and grape, including 35 amino acids and derivatives, 70 flavonoids, 141 lipids, 15 vitamins, and others. Rice harbored 238 up-regulated and 108 down-regulated metabolites (Figure 3c). We found 341 DAMs between corn and grape, consisting of 52 amino acids and derivatives, 66 flavonoids, 118 lipids, 22 vitamins, and others. Compared with those in grape, 258 and 83 metabolites accumulated with elevated and reduced levels in corn (Figure 3d). The 334 DAMs between wheat and grapes included 41 amino acids and derivatives, 70 flavonoids, 129 lipids, 15 vitamins, and others. Compared with grapes, wheat produced 231 and 103 metabolites with significantly higher and lower levels, respectively (Figure 3e). Compared with crops, DAMs of mango and banana were mainly amino acids and derivatives, flavonoids, lipids, and vitamins (Figure S1).

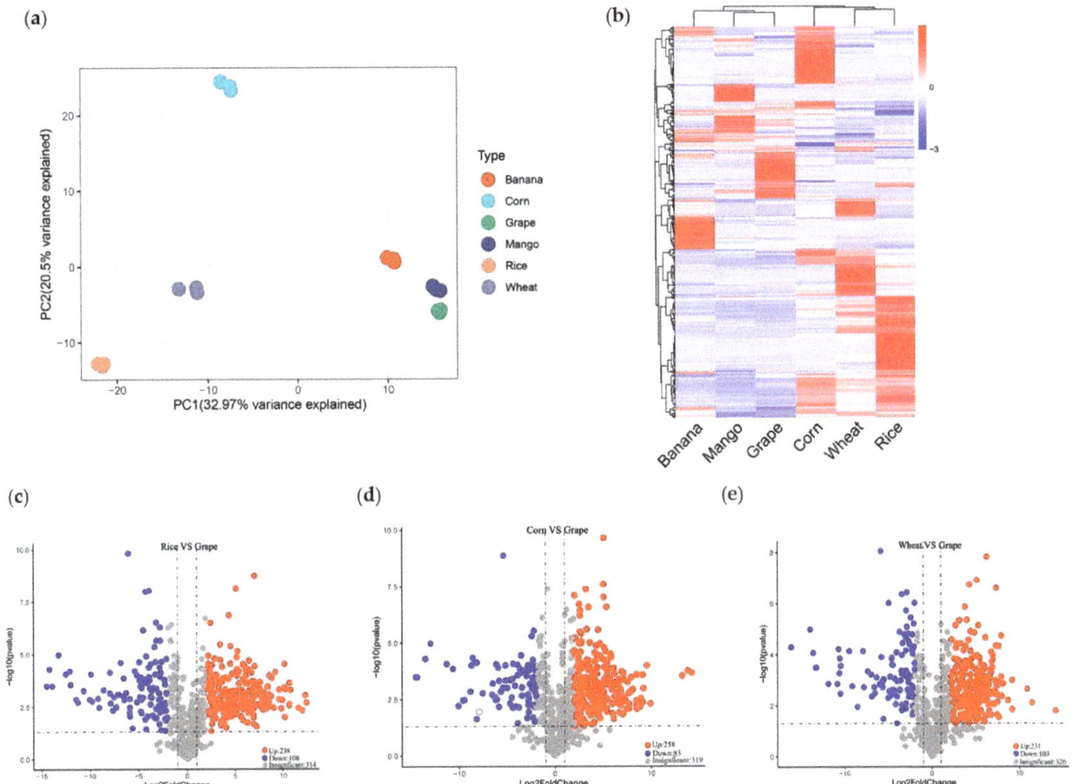

Figure 3. Q trap 6500+ LC-MS/MS was used to analyze metabolite changes in crops and fruits. (**a**) Principal component analysis of 660 metabolites detected in crops and fruits. (**b**) Heat map analysis of 664 metabolites detected in crops and fruits. (**c**) Volcanic map analysis of differentially accumulated metabolites in rice and grape. (**d**) Volcanic map analysis of differentially accumulated metabolites in corn and grape. (**e**) Volcanic map analysis of differentially accumulated metabolites in wheat and grape. The average of three biological replicates was used for metabolite analysis. The content of each metabolite was normalized, and hierarchical clustering was carried out. Each crop and fruit was labeled in a single column, and each metabolite was represented by a single row.

3.4. Comparative Analysis of Common Existing Nutrients in Crops and Fruits

To comprehend the nutritional value of crops and fruits, we firstly focused on common existing nutrients. The accumulation patterns of vitamins are complementary in crops and fruits. Corn was rich in most B vitamins, including tetrahydrofolate, nicotinamide, biotin, pyridoxal, pyridoxamine, and riboflavin. Rice and wheat accumulated the most vitamin B1 and phosphorylated B6, respectively (Figure 4a). Fruits contained vital compounds limited in crops. For instance, mango produced more vitamin C, and grape provided high content of 5-methyltetrahydrofolate. These vitamins are cofactors of many enzymes. Banana yielded a high level of non-phosphorylated vitamin B6 (Figure 4a). To conclude, the categories of vitamins between crops and fruits were significantly different. Crops were rich in B vitamins, which may exist in bran and corn endosperm. Fruits supplemented other necessary vitamins, such as vitamin C.

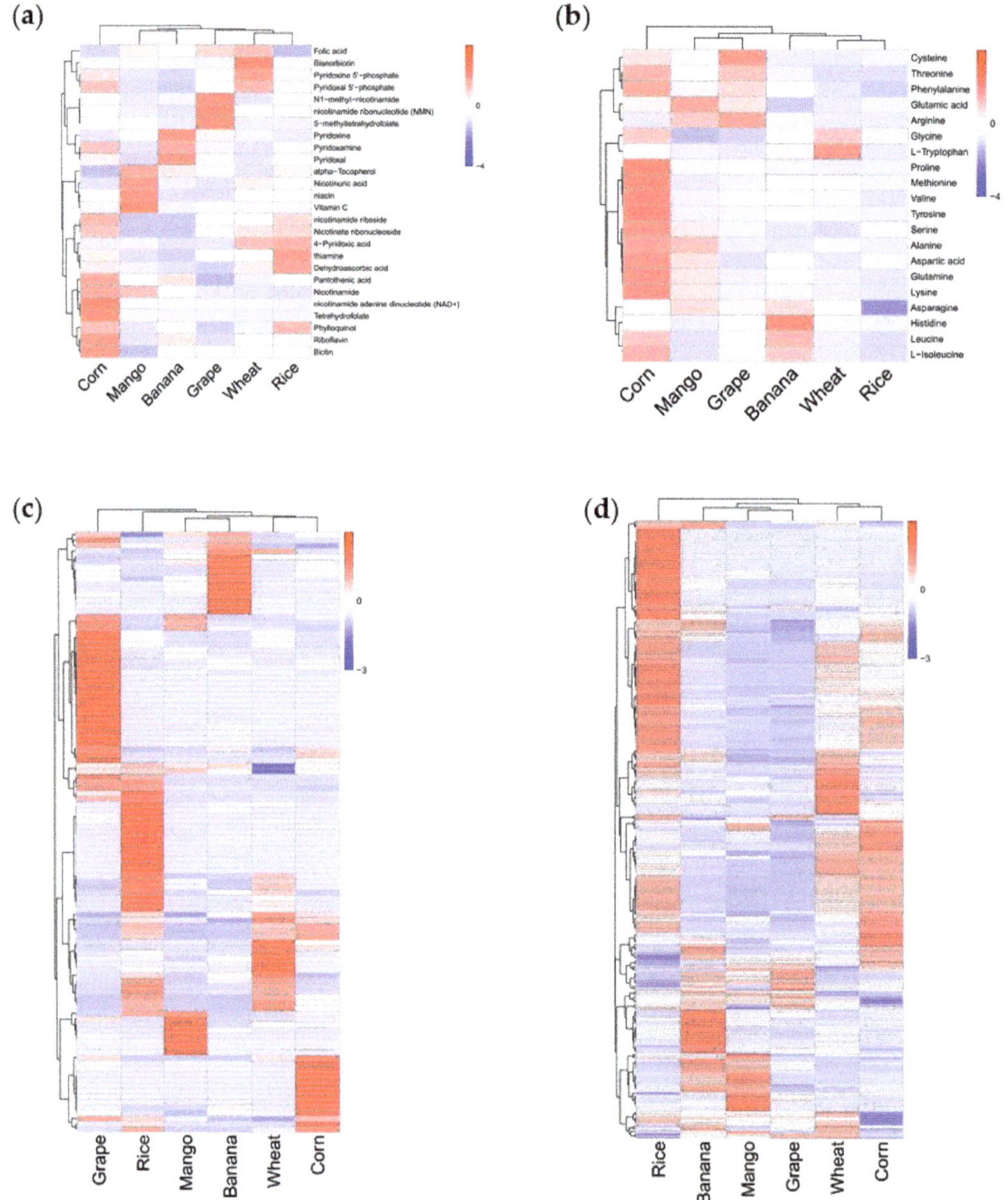

Figure 4. Accumulation of vitamins and amino acids in three crops and three fruits. (**a**) Heat map of vitamin in three crops and three fruits. (**b**) Heat map of amino acid in three crops and three fruits. (**c**) Heat map of flavonoids in three crops and three fruits. (**d**) Heat map of lipids in three crops and three fruits. The average of three biological replicates was used for metabolite analysis. The content of each metabolite was normalized, and hierarchical clustering was carried out. Each crop and fruit is marked in a single column, and each metabolite is represented by a single row.

The accumulation of amino acids was also different in crops and fruits. Corn yielded high levels of essential amino acids, excluding tryptophan, while wheat produced considerable content of tryptophan. Compared with grapes and banana, mango contains more amino acids (Figure 4b).

Lipids and flavonoids also differed in crops and fruits. Flavonoids are highly accumulated in grapes and rice, but less so in the other species (Figure 4c). The accumulation patterns across species were complementary. Flavanones and anthocyanins mainly existed in grapes, while flavones were mainly present in rice. Crops produced more lipids than fruits, especially more than grape (Figure 4d).

3.5. Comparative Analysis of Species-Specific Nutrients

Some nutrients showed remarkable species dependence. We focused on compounds commonly present in all fruits but absent in crops. We detected five fruit-specific flavonoids, including (-)-epigallocatechin, catechin, (+)-gallocatechin, (-)-gallatechin, and eriodictyol 7-O-glucoside. Meanwhile, crops harbored more kinds of specific metabolites, consisting of four flavonoids, three lipids, and four others (Table 1). Then, we noticed metabolites existing in a single species: One in wheat, twelve in rice, five in corn, five in mango, five in grape, six in banana (Table S5). In addition, we analyzed some crucial metabolites, such as chlorogenic acid and theanine. Chlorogenic acid was detected only in corn, grape, and banana. Although theanine was present in all species, it accumulated at higher levels in mango and grape (Table S5).

Table 1. Specific metabolites found in wheat, grape, and mango.

Species	ID	Q1 (Da)	RT (min)	Compounds	Class
	hsf030	301.1	5.63	Diosmetin	Flavonoid
	MP0307	301.05	8.04	Chrysoeriol	Flavonoid
	MP0566	565.1	5.64	C-pentosyl-apigenin O-hexoside	Flavonoid
	MP0577	581.1	4.98	C-pentosyl-luteolin O-hexoside	Flavonoid
	hsl047	325.3	12.49	Heneicosanoic acid (C21:0)	Lipids
Crops	MP0557	548.35	10.8	LysoPC 20:2	Lipids
	MP2054	307.3	11.39	cis-11,14,17-Eicosatrienoic Acid (C20:3)	Lipids
	hsb025	166.04	3.46	2-(Formylamino)benzoic acid	Others
	MP0885	287	12.14	5-deoxo-ent-10-oxodepressin	Others
	hsa126	161.2	0.72	D-Alanyl-D-Alanine	Amino acid and its derivatives
	MP0576	580.9	5.18	inositol pentakisphosphate	Organic acid and its derivates
	hsc303	307	3.5	(-)-Epigallocatechin	Flavonoid
	hsc317	291.08	3.58	Catechin	Flavonoid
Fruits	hsc325	307	2.57	(+)-Gallocatechin	Flavonoid
	hsc327	307	2.12	(-)-gallatechin	Flavonoid
	hsf421	451	3.72	Eriodictyol 7-O-glucoside	Flavonoid

4. Discussion

Crops and fruits are indispensable for humans. They provide a variety of nutrients and protect humans against diseases. Hence, nutritional metabolomics of crops and fruits are essential to dissect the nutritional value and maintain a healthy diet. Although metabolome studies have been conducted in crops and fruits, metabolic differences between crops and fruits are yet to be drawn. In this study, we detected 2631 and 660 compounds in crops and fruits using non-targeted and targeted LC-MS, respectively. Metabolic features differed remarkably in crops and fruits. Moreover, this work revealed the complementary pattern of nutrient accumulation in different species, which provides metabolic insights into food choice.

Plant metabolites play vital roles in plant growth and nutrition [31]. Tremendous metabolic diversity occurs across different species [32]. An investigation on metabolomic data has illustrated interspecific metabolic variation in rice and maize populations, and identified flavonoids and phenolamides as contributory compounds in diversified evolution in rice and maize [33]. A study with widely targeted LC-MS/MS has analyzed metabolic features of the pearling fractions and discovered the enrichment of health-beneficial metabolites in the wheat bran. Meanwhile, the authors have also found that flavonoids are of the most remarkable spatial divergence among grain layers [34]. In ten fruits, the diversity of more than 2000 metabolites has been studied and used to construct a metabolic evolutionary tree [32]. Through the analysis of metabolites in the fruits and leaves of 12 kinds of piper, it was found that, although there were significant differences in chemical richness and structural complexity between different species, fruit diversity exceeded leaf diversity [35]. Although these works have elucidated some of the metabolic diversity in different species, our work directly provides metabolic evidence for the necessity of a balance between staple foods and fruit uptake. In this study, widely targeted metabolomics were used to study the metabolic diversity in major food crops (rice, wheat, and corn) and three fruits (mango, grape, and banana). The main differential metabolites in crops and fruits were vitamins, amino acids, flavonoids, and lipids. They are essential in plant growth and development, as well as in keeping humans fit.

Common existing nutrients displayed remarkable species-dependent accumulation [36,37]. Independent work has documented accumulation patterns of flavonoids in rice and grape, respectively [33,38]. By using a comparative metabolomic analysis, we found that flavonoids were most abundant in rice and grape among the six species. Grape contains high contents of anthocyanins and 5-methyltetrahydrofolate. Moreover, our data show that mango is rich in vitamin C and vitamin E, which resembles a previous work [39]. Furthermore, we have found that the contents of most vitamins and amino acids in mango are the highest among the three fruits, while the content of vitamin B6 in banana was the highest. Compared with the three fruits, the content of lipid in the three crops was higher. Although staple crops provide energy for humans worldwide, they lack essential nutrients. Although sweet corn contains relatively high levels of most amino acids and vitamins, the content of vitamins and amino acids in rice and wheat were relatively low. Rice mainly contained vitamin B1, and wheat mainly contained vitamin B6 and tryptophan. Our work discovered the complementary pattern of essential nutrients in crops and fruits.

Bioactive polyphenols from fruits have significant antioxidant activity [40]. For instance, catechin protects us against several diseases induced by oxidative stress, such as cardiovascular disease and cancer [41]. However, the direct application of catechins in food was limited by the low stability and content [42]. In this study, we draw the diversity of catechin in fruits, which provides insights into food choice based on the abundance of health-beneficial compounds. In addition, we also detected resveratrol in grape. It has strong antioxidant properties and the effects of protecting the heart and blood vessels, anti-arrhythmia, and vasodilation [43]. We found 11 specific metabolites in crops, mainly from flavonoids and lipids. These specific metabolites enriched the diversity of metabolites in different species and met humans' needs for nutrients.

5. Conclusions

This work identified the metabolic diversity in three major staple crops and three fruits, revealed the complementary patterns of nutrient accumulation of different species, and decoded the species-specific patterns of bioactive compounds. Among the three crops, sweet corn was rich in vitamins and amino acids, while rice and wheat were deficient in vitamins and amino acids. Among the three fruits, mango was rich in vitamins and amino acids. Compared with fruits, crops were rich in lipids. Overall, this work provided metabolomic evidence for a healthy diet, which aims to highlight the need for macronutrients and essential micronutrients [44].

Supplementary Materials: The following supporting information can be downloaded at: https://www.mdpi.com/article/10.3390/foods11040550/s1, Table S1: Metabolic signals in the three fruits and three crops were detected by LC-MS-based non-targeted; Table S2: Metabolite variation in the three fruits and three crops were detected by LC-MS-based non-targeted; Table S3:Metabolites in the three fruits and three crops were detected by LC-MS-based targeted; Table S4: Specific metabolites in the three fruits and three crops were detected by LC-MS-based targeted; Table S5: Metabolite variation in the three fruits and three crops were detected by LC-MS-based targeted. Figure S1: Q trap 6500$^+$ LC-MS/MS was used to analyze metabolite changes in crops and fruits. (a) Volcanic map analysis of differentially accumulated metabolites in rice and banana. (b) Volcanic map analysis of differentially accumulated metabolites in corn and banana. (c) Volcanic map analysis of differentially accumulated metabolites in wheat and banana. (d) Volcanic map analysis of differentially accumulated metabolites in rice and mango. (e) Volcanic map analysis of differentially accumulated metabolites in corn and mango. (f) Volcanic map analysis of differentially accumulated metabolites in wheat and mango.

Author Contributions: Conceptualization, C.F.; software, M.L. and Y.W.; validation, Y.S., K.L. and Y.W.; formal analysis, Y.S. and Y.G.; investigation, Y.S., Y.W., M.L. and Y.G.; resources, C.F., X.L. and J.L.; data curation, Y.S. and X.L.; writing—original draft preparation, Y.S. and K.L.; writing—review and editing, Y.S. and C.F.; visualization, Y.S.; project administration, C.F.; funding acquisition, K.L. and J.L. All authors have read and agreed to the published version of the manuscript.

Funding: This research was funded by the Key Research and Development Program of Hainan (ZDYF2020066), the Natural Science Foundation of Hainan Province (321RC463), Hainan Academician Innovation Platform (HD-YSZX-202003 and HD-YSZX-202004), and the Hainan University Startup Fund (KYQD(ZR)1866).

Institutional Review Board Statement: Not applicable.

Informed Consent Statement: Not applicable.

Data Availability Statement: All data and materials are available on request.

Acknowledgments: Authors would like to thank the reviewers for their time, effort, insightful comments, and suggestions that helped improve this manuscript.

Conflicts of Interest: The authors declare no conflict of interest.

References

1. Hindu, V.; Palacios-Rojas, N.; Babu, R.; Suwarno, W.; Rashid, Z.; Usha, R.; Saykhedkar, G.; Nair, S. Identification and validation of genomic regions influencing kernel zinc and iron in maize. *Thero. Appl. Genet.* **2018**, *131*, 1443–1457. [CrossRef] [PubMed]
2. Lowe, N. The global challenge of hidden hunger: Perspectives from the field. *Proc. Nutr. Soc.* **2021**, *80*, 1–7. [CrossRef] [PubMed]
3. Gashu, D.; Nalivata, P.; Amede, T.; Ander, E.; Bailey, E.; Botoman, L.; Chagumaira, C.; Gameda, S.; Haefele, S.; Hailu, K.; et al. The nutritional quality of cereals varies geospatially in Ethiopia and Malawi. *Nature* **2021**, *594*, 71–76. [CrossRef] [PubMed]
4. Palacios-Rojas, N.; McCulley, L.; Kaeppler, M.; Titcomb, T.; Gunaratna, N.; Lopez-Ridaura, S.; Tanumihardjo, S. Mining maize diversity and improving its nutritional aspects within agro-food systems. *Compr. Rev. Food Sci. Food Saf.* **2020**, *19*, 1809–1834. [CrossRef] [PubMed]
5. Goredema-Matongera, N.; Ndhlela, T.; Magorokosho, C.; Kamutando, C.; van Biljon, A.; Labuschagne, M. Multinutrient Biofortification of Maize (*Zea mays* L.) in Africa: Current Status, Opportunities and Limitations. *Nutrients* **2021**, *13*, 1039. [CrossRef] [PubMed]
6. Rondanelli, M.; Miccono, A.; Peroni, G.; Nichetti, M.; Infantino, V.; Spadaccini, D.; Alalwan, T.; Faliva, M.; Perna, S. Rice germ macro- and micronutrients: A new opportunity for the nutraceutics. *Nat. Prod. Res.* **2021**, *35*, 1532–1536. [CrossRef]
7. Yu, S.; Tian, L. Breeding Major Cereal Grains through the Lens of Nutrition Sensitivity. *Mol. Plant* **2018**, *11*, 23–30. [CrossRef]
8. Liu, R. Potential synergy of phytochemicals in cancer prevention: Mechanism of action. *J. Nutr.* **2004**, *134*, 3479S–3485S. [CrossRef]
9. Liu, R. Dietary bioactive compounds and their health implications. *J. Food Sci.* **2013**, *78*, A18–A25. [CrossRef]
10. Brookie, K.; Best, G.; Conner, T. Intake of Raw Fruits and Vegetables Is Associated with Better Mental Health Than Intake of Processed Fruits and Vegetables. *Front. Psychol.* **2018**, *9*, 487. [CrossRef]
11. Duthie, S.; Duthie, G.; Russell, W.; Kyle, J.; Macdiarmid, J.; Rungapamestry, V.; Stephen, S.; Megias-Baeza, C.; Kaniewska, J.; Shaw, L.; et al. Effect of increasing fruit and vegetable intake by dietary intervention on nutritional biomarkers and attitudes to dietary change: A randomised trial. *Eur. J. Nutr.* **2018**, *57*, 1855–1872. [CrossRef] [PubMed]

12. Ediriweera, M.; Tennekoon, K.; Samarakoon, S. A Review on Ethnopharmacological Applications, Pharmacological Activities, and Bioactive Compounds of *Mangifera indica* (Mango). *Evid. Based Complement Altern. Med.* **2017**, *2017*, 6949835. [CrossRef]
13. Ruocco, S.; Stefanini, M.; Stanstrup, J.; Perenzoni, D.; Mattivi, F.; Vrhovsek, U. The metabolomic profile of red non-V. vinifera genotypes. *Food Res. Int. (Ott. Ont.)* **2017**, *98*, 10–19. [CrossRef] [PubMed]
14. Maldonado-Celis, M.; Yahia, E.; Bedoya, R.; Landázuri, P.; Loango, N.; Aguillón, J.; Restrepo, B.; Guerrero Ospina, J. Chemical Composition of Mango (*Mangifera indica* L.) Fruit: Nutritional and Phytochemical Compounds. *Front. Plant Sci.* **2019**, *10*, 1073. [CrossRef]
15. Lebaka, V.; Wee, Y.; Ye, W.; Korivi, M. Nutritional Composition and Bioactive Compounds in Three Different Parts of Mango Fruit. *Int. J. Environ. Res. Public Health* **2021**, *18*, 741. [CrossRef] [PubMed]
16. Mwaurah, P.; Kumar, S.; Kumar, N.; Panghal, A.; Attkan, A.; Singh, V.; Garg, M. Physicochemical characteristics, bioactive compounds and industrial applications of mango kernel and its products: A review. *Compr. Rev. Food Sci. Food Saf.* **2020**, *19*, 2421–2446. [CrossRef]
17. Fu, X.; Cheng, S.; Feng, C.; Kang, M.; Huang, B.; Jiang, Y.; Duan, X.; Grierson, D.; Yang, Z. Lycopene cyclases determine high α-/β-carotene ratio and increased carotenoids in bananas ripening at high temperatures. *Food Chem.* **2019**, *283*, 131–140. [CrossRef]
18. Liu, Q.; Tang, G.; Zhao, C.; Feng, X.; Xu, X.; Cao, S.; Meng, X.; Li, S.; Gan, R.; Li, H. Comparison of Antioxidant Activities of Different Grape Varieties. *Molecules* **2018**, *23*, 2432. [CrossRef]
19. Das, S.; Laskar, M.; Sarker, S.; Choudhury, M.; Choudhury, P.; Mitra, A.; Jamil, S.; Lathiff, S.; Abdullah, S.; Basar, N.; et al. Prediction of Anti-Alzheimer's Activity of Flavonoids Targeting Acetylcholinesterase in silico. *Phytochem. Anal.* **2017**, *28*, 324–331. [CrossRef]
20. Chen, J.; Hu, X.; Shi, T.; Yin, H.; Sun, D.; Hao, Y.; Xia, X.; Luo, J.; Fernie, A.; He, Z.; et al. Metabolite-based genome-wide association study enables dissection of the flavonoid decoration pathway of wheat kernels. *Plant Biotechnol. J.* **2020**, *18*, 1722–1735. [CrossRef]
21. Xu, G.; Cao, J.; Wang, X.; Chen, Q.; Jin, W.; Li, Z.; Tian, F. Evolutionary Metabolomics Identifies Substantial Metabolic Divergence between Maize and Its Wild Ancestor, Teosinte. *Plant Cell* **2019**, *31*, 1990–2009. [CrossRef] [PubMed]
22. Peng, M.; Shahzad, R.; Gul, A.; Subthain, H.; Shen, S.; Lei, L.; Zheng, Z.; Zhou, J.; Lu, D.; Wang, S.; et al. Differentially evolved glucosyltransferases determine natural variation of rice flavone accumulation and UV-tolerance. *Nat. Commun.* **2017**, *8*, 1975. [CrossRef] [PubMed]
23. Shi, T.; Zhu, A.; Jia, J.; Hu, X.; Chen, J.; Liu, W.; Ren, X.; Sun, D.; Fernie, A.; Cui, F.; et al. Metabolomics analysis and metabolite-agronomic trait associations using kernels of wheat (*Triticum aestivum*) recombinant inbred lines. *Plant J.* **2020**, *103*, 279–292. [CrossRef] [PubMed]
24. Wen, W.; Jin, M.; Li, K.; Liu, H.; Xiao, Y.; Zhao, M.; Alseekh, S.; Li, W.; de Abreu, E.; Lima, F.; et al. An integrated multi-layered analysis of the metabolic networks of different tissues uncovers key genetic components of primary metabolism in maize. *Plant J.* **2018**, *93*, 1116–1128. [CrossRef]
25. Dong, X.; Gao, Y.; Chen, W.; Wang, W.; Gong, L.; Liu, X.; Luo, J. Spatio-temporal distribution of phenolamides and the genetics of natural variation of hydroxycinnamoyl spermidine in rice. *Mol. Plant* **2015**, *8*, 111–121. [CrossRef]
26. Peng, Z.; Zhang, H.; Li, W.; Yuan, Z.; Xie, Z.; Zhang, H.; Cheng, Y.; Chen, J.; Xu, J. Comparative profiling and natural variation of polymethoxylated flavones in various citrus germplasms. *Food Chem.* **2021**, *354*, 129499. [CrossRef]
27. Lombardo, V.; Osorio, S.; Borsani, J.; Lauxmann, M.; Bustamante, C.; Budde, C.; Andreo, C.; Lara, M.; Fernie, A.; Drincovich, M. Metabolic profiling during peach fruit development and ripening reveals the metabolic networks that underpin each developmental stage. *Mol. Plant* **2011**, *157*, 1696–1710. [CrossRef]
28. Chen, W.; Gong, L.; Guo, Z.; Wang, W.; Zhang, H.; Liu, X.; Yu, S.; Xiong, L.; Luo, J. A novel integrated method for large-scale detection, identification, and quantification of widely targeted metabolites: Application in the study of rice metabolomics. *Mol. Plant* **2013**, *6*, 1769–1780. [CrossRef]
29. Horai, H.; Arita, M.; Kanaya, S.; Nihei, Y.; Ikeda, T.; Suwa, K.; Ojima, Y.; Tanaka, K.; Tanaka, S.; Aoshima, K.; et al. MassBank: A public repository for sharing mass spectral data for life sciences. *J. Proteome Res.* **2010**, *45*, 703–714. [CrossRef]
30. Wishart, D.; Tzur, D.; Knox, C.; Eisner, R.; Guo, A.; Young, N.; Cheng, D.; Jewell, K.; Arndt, D.; Sawhney, S.; et al. HMDB: The Human Metabolome Database. *Nucleic Acids Res.* **2007**, *35*, D521–D526. [CrossRef]
31. Chen, W.; Wang, W.; Peng, M.; Gong, L.; Gao, Y.; Wan, J.; Wang, S.; Shi, L.; Zhou, B.; Li, Z.; et al. Comparative and parallel genome-wide association studies for metabolic and agronomic traits in cereals. *Nat. Commun.* **2016**, *7*, 12767. [CrossRef] [PubMed]
32. Qi, J.; Li, K.; Shi, Y.; Li, Y.; Dong, L.; Liu, L.; Li, M.; Ren, H.; Liu, X.; Fang, C.; et al. Cross-Species Comparison of Metabolomics to Decipher the Metabolic Diversity in Ten Fruits. *Metabolites* **2021**, *11*, 164. [CrossRef]
33. Deng, M.; Zhang, X.; Luo, J.; Liu, H.; Wen, W.; Luo, H.; Yan, J.; Xiao, Y. Metabolomics analysis reveals differences in evolution between maize and rice. *Plant J.* **2020**, *103*, 1710–1722. [CrossRef] [PubMed]
34. Zhu, A.; Zhou, Q.; Hu, S.; Wang, F.; Tian, Z.; Hu, X.; Liu, H.; Jiang, D.; Chen, W. Metabolomic analysis of the grain pearling fractions of six bread wheat varieties. *Food Chem.* **2022**, *369*, 130881. [CrossRef] [PubMed]
35. Schneider, G.; Salazar, D.; Hildreth, S.; Helm, R.; Whitehead, S. Comparative Metabolomics of Fruits and Leaves in a Hyperdiverse Lineage Suggests Fruits Are a Key Incubator of Phytochemical Diversification. *Front Plant Sci.* **2021**, *12*, 693739. [CrossRef] [PubMed]
36. Hectors, K.; Van Oevelen, S.; Geuns, J.; Guisez, Y.; Jansen, M.; Prinsen, E. Dynamic changes in plant secondary metabolites during UV acclimation in *Arabidopsis thaliana*. *Physiol. Plant.* **2014**, *152*, 219–230. [CrossRef]

37. Fang, C.; Luo, J. Metabolic GWAS-based dissection of genetic bases underlying the diversity of plant metabolism. *Plant J.* **2019**, *97*, 91–100. [CrossRef]
38. Pérez-Navarro, J.; Izquierdo-Cañas, P.; Mena-Morales, A.; Martínez-Gascueña, J.; Chacón-Vozmediano, J.; García-Romero, E.; Hermosín-Gutiérrez, I.; Gómez-Alonso, S. Phenolic compounds profile of different berry parts from novel Vitis vinifera L. red grape genotypes and Tempranillo using HPLC-DAD-ESI-MS/MS: A varietal differentiation tool. *Food Chem.* **2019**, *295*, 350–360. [CrossRef]
39. Barbosa Gámez, I.; Caballero Montoya, K.; Ledesma, N.; Sáyago Ayerdi, S.; García Magaña, M.; Bishop von Wettberg, E.; Montalvo-González, E. Changes in the nutritional quality of five *Mangifera* species harvested at two maturity stages. *J. Sci. Food Agric.* **2017**, *97*, 4987–4994. [CrossRef]
40. Tan, L.; Jin, Z.; Ge, Y.; Nadeem, H.; Cheng, Z.; Azeem, F.; Zhan, R. Comprehensive ESI-Q TRAP-MS/MS based characterization of metabolome of two mango (*Mangifera indica* L.) cultivars from China. *Sci. Rep.* **2020**, *10*, 20017. [CrossRef]
41. Ma, W.; Waffo-Téguo, P.; Jourdes, M.; Li, H.; Teissedre, P. First evidence of epicatechin vanillate in grape seed and red wine. *Food Chem.* **2018**, *259*, 304–310. [CrossRef] [PubMed]
42. Sabaghi, M.; Hoseyni, S.; Tavasoli, S.; Mozafari, M.; Katouzian, I. Strategies of confining green tea catechin compounds in nano-biopolymeric matrices: A review. *Colloids Surf. B Biointerfaces* **2021**, *204*, 111781. [CrossRef] [PubMed]
43. Xia, E.; Deng, G.; Guo, Y.; Li, H. Biological activities of polyphenols from grapes. *Int. J. Mol. Sci.* **2010**, *11*, 622–646. [CrossRef] [PubMed]
44. FAO; IFAD; UNICEF; WFP; WHO. *The State of Food Security and Nutrition in the World 2021.Transforming food Systems for Food Security, Improved Nutrition and Affordable Healthy Diets for All*; FAO: Rome, Italy, 2021. [CrossRef]

Article

Integrated Metabolomics and Volatolomics for Comparative Evaluation of Fermented Soy Products

Sang-Hee Lee [1], Sunmin Lee [1], Seung-Hwa Lee [2], Hae-Jin Kim [2], Digar Singh [1] and Choong-Hwan Lee [1,*]

[1] Department of Bioscience and Biotechnology, Konkuk University, Seoul 05029, Korea; tkdgml823@hanmail.net (S.-H.L.); duly123@naver.com (S.L.); singhdigar@gmail.com (D.S.)
[2] Experiment Research Institute, National Agricultural Products Quality Management Service, Gimcheon-si 39660, Korea; shlee96@korea.kr (S.-H.L.); asarela00@korea.kr (H.-J.K.)
* Correspondence: chlee123@konkuk.ac.kr

Abstract: Though varying metabolomes are believed to influence distinctive characteristics of different soy foods, an in-depth, comprehensive analysis of both soluble and volatile metabolites is largely unreported. The metabolite profiles of different soy products, including cheonggukjang, meju, doenjang, and raw soybean, were characterized using LC-MS (liquid chromatography–mass spectrometry), GC-MS (gas chromatography–mass spectrometry), and headspace solid-phase microextraction (HS-SPME) GC-MS. Principal component analysis (PCA) showed that the datasets for the cheonggukjang, meju, and doenjang extracts were distinguished from the non-fermented soybean across PC1, while those for cheonggukjang and doenjang were separated across PC2. Volatile organic compound (VOC) profiles were clearly distinct among doenjang and soybean, cheonggukjang, and meju samples. Notably, the relative contents of the isoflavone glycosides and DDMP (2,3-dihydro-2,5-dihydroxy-6-methyl-4H-pyran-4-one) conjugated soyasaponins were higher in soybean and cheonggukjang, compared to doenjang, while the isoflavone aglycones, non-DDMP conjugated soyasaponins, and amino acids were significantly higher in doenjang. Most VOCs, including the sulfur containing compounds aldehydes, esters, and furans, were relatively abundant in doenjang. However, pyrazines, 3-methylbutanoic acid, maltol, and methoxyphenol were higher in cheonggukjang, which contributed to the characteristic aroma of soy foods. We believe that this study provides the fundamental insights on soy food metabolomes, which determine their nutritional, functional, organoleptic, and aroma characteristics.

Keywords: fermented soy product; metabolomics; volatolomics; metabolic pathway

Citation: Lee, S.-H.; Lee, S.; Lee, S.-H.; Kim, H.-J.; Singh, D.; Lee, C.-H. Integrated Metabolomics and Volatolomics for Comparative Evaluation of Fermented Soy Products. *Foods* **2021**, *10*, 2516. https://doi.org/10.3390/foods10112516

Academic Editors: Yelko Rodríguez Carrasco and Bojan Šarkanj

Received: 24 September 2021
Accepted: 18 October 2021
Published: 20 October 2021

Publisher's Note: MDPI stays neutral with regard to jurisdictional claims in published maps and institutional affiliations.

1. Introduction

Fermentation is a metabolic bioprocess that causes biochemical changes in the organic substrate, through the action of microbial enzymes [1]. Fermentation is considered a customarily important method of food processing, as it adds superior taste, flavor, and nutrition to the end-product, as well as enhanced shelf life, compared to raw materials [2]. First cultivated in East Asia, thousands of years ago, soybean is an important, nutrient-rich crop, but the presence of anti-nutritive toxic substances may limit its nutritional level [3]. Thus, the fermentation approach was used to overcome these limitations and improve the quality of soybeans, such as the desired taste, aroma, and health effects.

Depending on the fermentation process, various fermented foods are produced, such as cheonggukjang (fermented soybean), meju (fermented soybean block), doenjang (fermented soybean paste), ganjang (fermented soy sauce), and gochujang (fermented hot-pepper paste) [4]. Among them, the cheonggukjang, prepared traditionally, is made by steaming soybeans for several days with rice straw, and meju is made from boiled beans, wrapped in rice straw and fermented for several months. On the other hand, doenjang is made by fermenting the meju for an extended duration, ranging 3 months to 3 years, and separating the liquid layer as ganjang (Figure 1) [5]. Different fermented food products have varying flavors and nutritional functions, owing to the varying substrate processing and fermentation methods. However, the exact factors affecting the characteristics of each product are not precisely known.

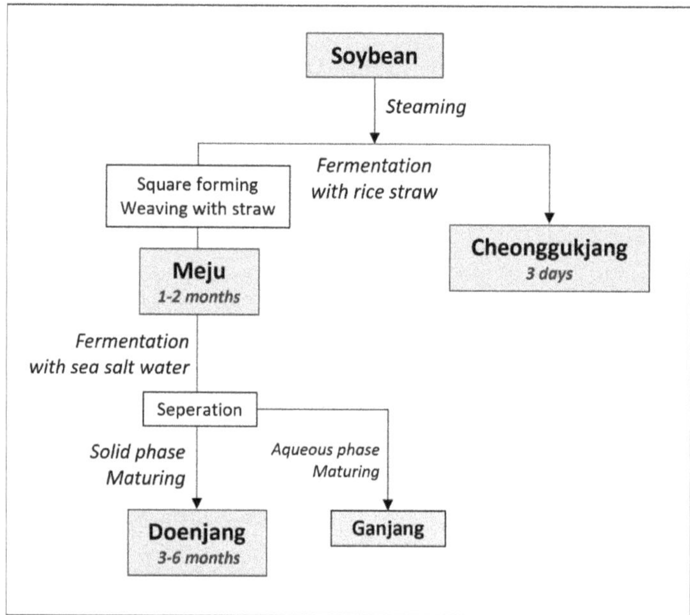

Figure 1. Manufacturing steps during the fermentative manufacturing of cheonggukjang, meju, and doenjang from raw soybean.

Metabolite profiling has been used in previous studies to assess the metabolomes of different foods and biological systems, using a variety of hyphenated mass spectrometry (MS)-based platforms [4,6]. Untargeted metabolomic studies, involving soybean fermented foods, can unravel the biochemical aspects of associated nutritional and functional activities and may further reveal the biomarkers for their quality standardization.

The composition of volatile organic compounds, part of secondary metabolites in food, affects the taste, aroma, and properties, and, especially in fermented foods, it is considered an important criterion of consumer acceptability and require comprehensive analysis. SPME (solid phase microextraction) is a simple, solvent-free, and reliable microextraction technique to extract volatile compounds from solids, liquids, or even gasses [7,8]. It is considered a highly reproducible method and has been employed in food aroma and perfumery studies. This can detect volatile compounds that are highly unstable and difficult to detect in solvent-based extraction methods. Therefore, a combination of different spectroscopic platforms, such as GC–MS, LC–MS, and SPME–GC–MS, could detect broader classes of metabolites and volatile compounds.

A recent study described the effect of fermentation methods on changes in total metabolite content by comparing various soy fermented products [9]. However, soybean differs in nutritional content and characteristics, depending on cultured area and cultivar. This can be a biased factor when comparing the differences in metabolite content by the fermentation method. Within our literature search, the comparison of volatile compounds for different soy fermented foods has not yet been carried out. Thus, comparison of primary and secondary metabolites, as well as the volatile organic compounds (VOCs) for different commercially available fermented soy products, including cheonggukjang, meju, and doenjang, which were produced traditionally, was performed. Additionally, soybean, which is used as raw material in most fermented soy foods, was analyzed.

2. Materials and Methods

2.1. Chemicals and Reagents

HPLC-grade water, acetonitrile, and methanol were purchased from Fisher Scientific (Pittsburgh, PA, USA). Diethylene glycol was obtained from Junsei Chemical Co., Ltd. (Tokyo, Japan). Formic acid, methoxyamine hydrochloride, pyridine, potassium persulfate, N-methyl-N-(trimethylsilyl) trifluoroacetamide (MSTFA), 2,2′-azino-bis (3-ethylbenzothiazoline-6-sulfonic acid), diammonium salt (ABTS), 6-hydroxy-2,5,7,8-tetramethylchromane-2-carboxylic acid (Trolox), sodium acetate, acetic acid, 2,4,6-Tris (2-pyridyl)-s-triazine (TPTZ), hydrochloric acid (HCl), iron (III) chloride hexahydrate, sodium carbonate, Folin–Ciocalteu's phenol, sodium hydroxide, gallic acid, and naringin were obtained from Sigma-Aldrich (St. Louis, MO, USA).

2.2. Sample Information

Twenty samples, (5 soybean and 15 soybean fermented foods (5 cheonggukjang, 5 meju, and 5 doenjang)) were purchased from the traditional manufacturers (Supplementary Materials Table S1). Five different types of soybeans, cheonggukjang, meju, and doenjang were obtained from each of the five manufacturers, in order to obtain soybean fermented foods made from the same soybeans. All the samples were stored at $-20\,°C$, until analysis. Sample information and manufacturing steps are shown in Table S1 and Figure 1. The fermentation period was estimated based on a previous study [5].

2.3. Sample Extraction

2.3.1. Extraction and Derivatization for Metabolite Profiling

Sample extraction was carried out, following the methodology used by Lee et al. [10], with some modifications. Approximately 100 mg of sample powder was extracted with 80% aqueous methanol (100 mg/mL), using a mixer mill (Retsch GmbH and Co, Haan, Germany) for 10 min at 30 Hz/s, after sonication for 5 min. The sample mixtures were then centrifuged at $8000\times g$ for 10 min at $4\,°C$. The supernatant was separated, following the same procedure, repeated twice. All acquired samples were filtered through a 0.2 μm syringe filter, then dried using a speed vacuum concentrator (Biotron, Seoul, Korea). Because of the significantly different extraction yields, owing to the different physicochemical characteristics of samples, all extracts were normalized based on weight of samples (Table S2). The collected samples were resuspended in 80% methanol prior to the LC-MS and GC-MS procedures.

For the GC-MS analysis, sample extracts were added with methoxyamine hydrochloride (20 mg/mL) in pyridine and oximated at $30\,°C$ for 90 min. Then, the oximated samples were silylated with MSTFA at $37\,°C$ for 30 min. We maintained three replicates for fermented food sources and three analytical replicates.

2.3.2. Extraction of Headspace Volatile Organic Compounds

Extraction of volatile compounds was carried out by modifying the method by Jo et al. [11]. Sample (2.5 g) was mixed with 2.75 mL of water, 0.75 g of NaCl, and 10 μL of linalool (2500 ppm in distilled water), an internal standard. The sample mixture was added in the 20 mL amber SPME vials with a silicon/teflon septum (Supelco, Bellefonte, PA, USA). The mixtures were maintained at $60\,°C$ for 30 min, and SPME fiber (1 cm), coated with 75 μm carboxen/polydimethylsiloxane/divinylbenzene (CAR/PDMS/DVB), was exposed at $40\,°C$ for 30 min to the headspace. The fiber had a desorption procedure at $230\,°C$ for 5 min in a GC injector, and the analysis was performed in splitless mode.

2.4. Instrumentations

An Agilent 7890A GC system (Agilent Technologies, Palo Alto, CA, USA), comprising an Agilent 7693 autosampler, coupled with a TOF Pegasus III mass spectrometer (LECO, St. Joseph, MI, USA), was used in gas chromatography/time-of-flight mass spectrometry (GC–TOF-MS) analysis. An RTX-5MS column (30 m × 0.25 mm, 0.25 μm particle size; Restek Corp., Bellefonte, PA, USA) was used, with a flow of 1.5 mL/min of helium. Sample was injected with split ratio of 10:1. The oven temperatures, maintained at 75 °C for 2 min, increased to 300 °C at 15 °C/min, and then were maintained for 3 min. Details of the analysis and instrument operation conditions were adopted from the method suggested by Jung et al. [12].

Ultra-high performance liquid chromatography linear trap quadrupole tandem mass spectrometry (UHPLC-LTQ-orbitrap-MS/MS) was performed on an Vanquish Binary pump H system (Thermo Fisher Scientific, Waltham, MA, USA), equipped with auto-sampler. The column was an Phenomenex KINETEX® C18 column (100 mm × 2.1 mm, 1.7 μm particle size; Torrance, CA, USA), and the injection volume was 5 μL. The column temperature was set to 40 °C, and the flow rate was 0.3 mL/min. The 0.1% (v/v) of formic acid in water (mobile phase A) and 0.1% (v/v) of formic acid in acetonitrile (mobile phase B) were used, respectively. The MS data were collected in the range of 100–2000 m/z using an ion trap mass spectrometer (Thermo Fisher Scientific, Waltham, MA, USA). The analytical program, applied for the sample, was adopted from the method described by Kwon et al. [13].

Headspace solid-phase microextraction-gas chromatography–mass spectrometry (HS-SPME/GC-TOF-MS) analysis was conducted. DB-FFAP column (30 m length × 0.25 mm internal diameter × 0.25 μm film thickness, J&W Scientific, Folsom, CA, USA) was used to separate a sample analyte. The oven temperature was initially set at 40 °C for 6 min, increased to 200 °C at a rate of 4 °C/min, and then maintained at 200 °C for 5 min. The flow rate of a carrier gas, helium, was 0.8 mL/min. A mass scan range of 50–400 m/z and 70 eV of ionization energy were set.

2.5. Data Processing and Multivariate Statistical Analysis

GC-TOF-MS and HS-SPME/GC-TOF-MS raw data files were converted to a network common data form (NetCDF) (*.cdf) with the ChromaTOF software (LECO, St. Joseph, MI, USA). The UHPLC-LTQ-Orbitrap-MS raw data were acquired and converted into netCDF (*.cdf) using Xcalibur software (version 2.1, Thermo Fisher Scientific). Both CDF data were processed with the MetAlign software (http://www.metalign.nl, (accessed on 16 September 2021)) for data alignment, in accordance with peak mass (m/z) and retention time (min). Multivariate statistical analysis was processed using SIMCA-P+ 12.0 software (Umetrics; Umeå, Sweden). Principal component analysis (PCA) and partial least squares discriminant analysis (PLS-DA) were used to compare metabolic differences of samples. The significantly discriminant metabolites with a VIP value > 1.0 and p-value < 0.05 were selected using the PLS-DA model. The selected metabolites were tentatively identified by comparison of various analysis data, such as molecular weights, formula, retention time, mass fragment patterns, and mass spectrum of standard compounds, as well as the published references, the chemical dictionary version 7.2 (Chapman and Hall/CRC), in-house library (off-line database in laboratory made by analyzing standards), commercial databases, such as National Institutes of Standards and Technology (NIST) Library (version 2.0, 2011, FairCom, Gaithersburg, MD, USA), and the Human Metabolome Database (HMDB; http://www.hmdb.ca/, (accessed on 4 September 2021)). Significance ($p < 0.05$) was tested by employing the one-way ANOVA and Student's t-test, using predictive analytics software (PASW), Statistics 18 (SPSS Inc., Chicago, IL, USA).

2.6. Determination of Total Phenolics, Flavonoids, and Antioxidant Properties

The evaluation of antioxidant activity and bioactive compounds in fermented soy foods were performed using 2,2-azino-bis-(3-ethylbenzothiazoline-6-sulfonic acid) diammonium salt (ABTS), ferric reducing antioxidant power (FRAP), total flavonoid contents (TPC), and total flavonoid contents (TFC) assay. The methods were adopted from those conducted by Jung et al. [14], with modifications. All experiments maintained three replicates of each sample's extracts.

3. Results

3.1. Multivariate Statistical Analyses of Different Fermented Soy Products

Primary, secondary metabolites and VOCs profiles from cheonggukjang, meju, doenjang, and unfermented soybean were examined using GC–TOF-MS, UHPLC-LTQ-Orbitrap-MS, and HS-SPME/GC-TOF-MS. Variations in metabolite profiles, among raw soybean and fermented soy foods, were examined based on the multivariate analysis of the respective datasets. PCA and PLS-DA models, obtained from GC–TOF-MS analysis, displayed cheonggukjang, meju, and doenjang extracts that were clearly segregated from the unfermented soybean samples across PC1, which suggests their varying metabolite profiles, according to different fermentative bioprocesses. GC-MS datasets for fermented soy foods, meju, and doenjang were clustered separately from cheonggukjang and soybean across PC2, 10.2% (Figure 2A). However, the datasets for each fermented samples were clustered separately from unfermented soybean across PC1, 16.8%. Similar patterns between the datasets for fermented soy foods and soybean were evident in the corresponding PLS-DA plots (Figure 2B). In the PLS-DA plot, the stated satisfaction values of the X and Y variables were 0.324 (R2X) and 0.982 (R2Y), respectively, with a prediction accuracy of 0.970 (Q2).

LC-MS-based PCA and PLS-DA score plots for the secondary metabolite profiles of the fermented soy foods displayed similar patterns as those obtained for GC-MS (Figure 2C,D). Both the PCA and PLS-DA plots highlighted a marked segregation between the fermented soy foods (cheonggukjang, doenjang, and meju) and unfermented soybean across PC1 (14.1%) and PLS1 (14.1%), respectively. LC-MS datasets for meju and doenjang were clustered separately from those for cheonggukjang and soybean across PC2 (11.4%) and PLS2 (11.4%), respectively, in corresponding PCA and PLS-DA plots. In the PLS-DA plot, X and Y variables were 0.352 (R2X), 0.975 (R2Y), and 0.965 (Q2), respectively.

PCA and PLS-DA score plots for the VOCs profiles, based on the HS-SPME/-GC-TOF-MS datasets, indicated distinct patterns, compared to those obtained for primary and secondary metabolites (Figure 2E,F). Most notably, the PCA plot highlighted the clearly segregated VOC datasets obtained for doenjang from rest of the samples (cheonggukjang, meju, and soybean) across PC1 (9.79%), which suggests its distinctive VOCs composition, compared to others. The PLS-DA model displayed similar patterns as PCA model, with statistical variants 0.279 (R2X), 0.979 (R2Y), and 0.908 (Q2).

3.2. Significantly Discriminant Metabolites among the Fermented Soy Products

3.2.1. Primary Metabolite Profiling of Soybean and Its Fermented Foods

Significantly discriminant primary metabolites were selected at a VIP value > 1.0 and *p*-value < 0.05, using the PLS-DA model based on GC-TOF-MS analysis. As shown in Figure 3, the primary metabolites levels were markedly varied among different soy foods made using different fermentation methods. A total of 40 metabolites, including 15 amino acids, 5 carbohydrates, 10 organic acids, 4 lipids, and 4 nucleotides, were selected (Table S3). Notably, all amino acids, as well as most fatty and organic acids, displayed higher relative abundance in doenjang. However, the organic acids associated with the TCA (tricarboxylic acid cycle) cycle, such as citric and malic acid, showed higher relative abundance in unfermented soybean, while succinic and malonic acid were relatively higher in cheonggukjang. Nucleotide compounds were relatively more abundant in cheonggukjang.

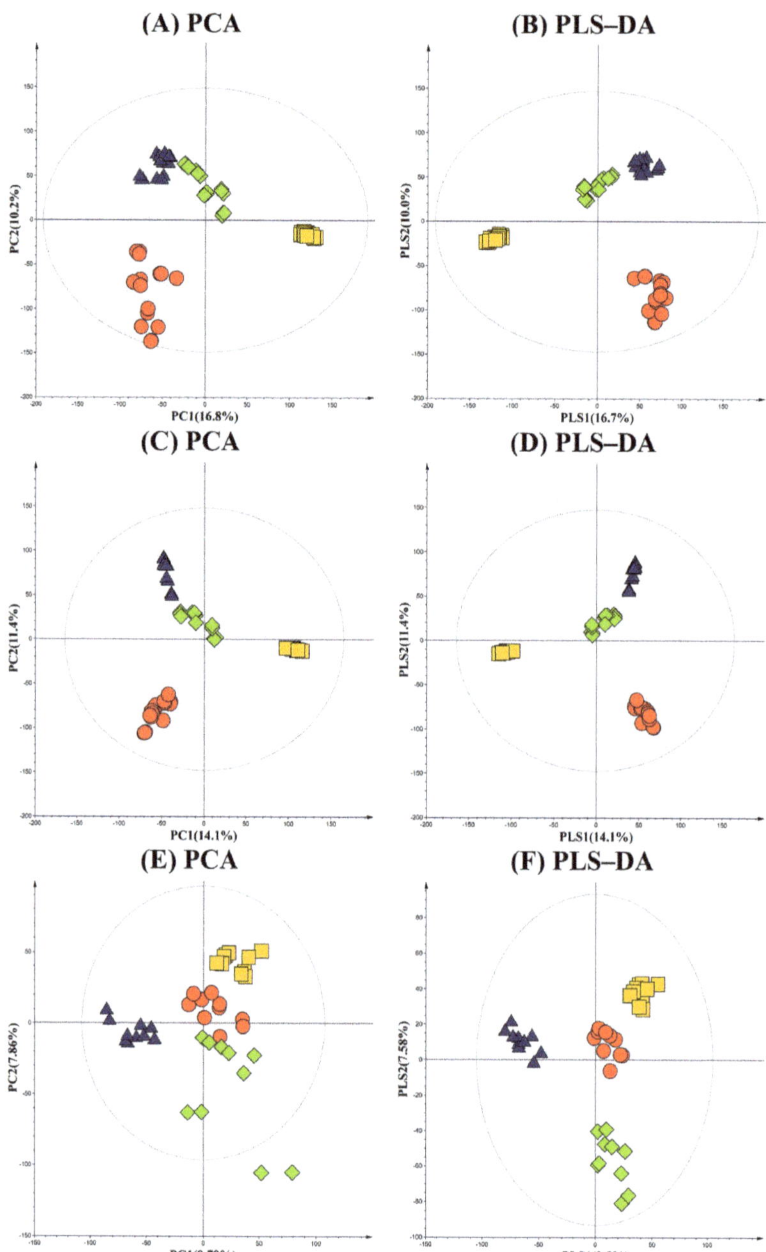

Figure 2. PCA and PLS-DA score plots, derived from the (**A**,**B**) GC-TOF-MS; (**C**,**D**) UHPLC-LTQ-Orbitrap-MS dataset; and (**E**,**F**) HS-SPME/GC-TOF-MS. The datasets are indicated - ▮; soy, - ●; cheonggukjang, ◆; meju, ▲; doenjang.

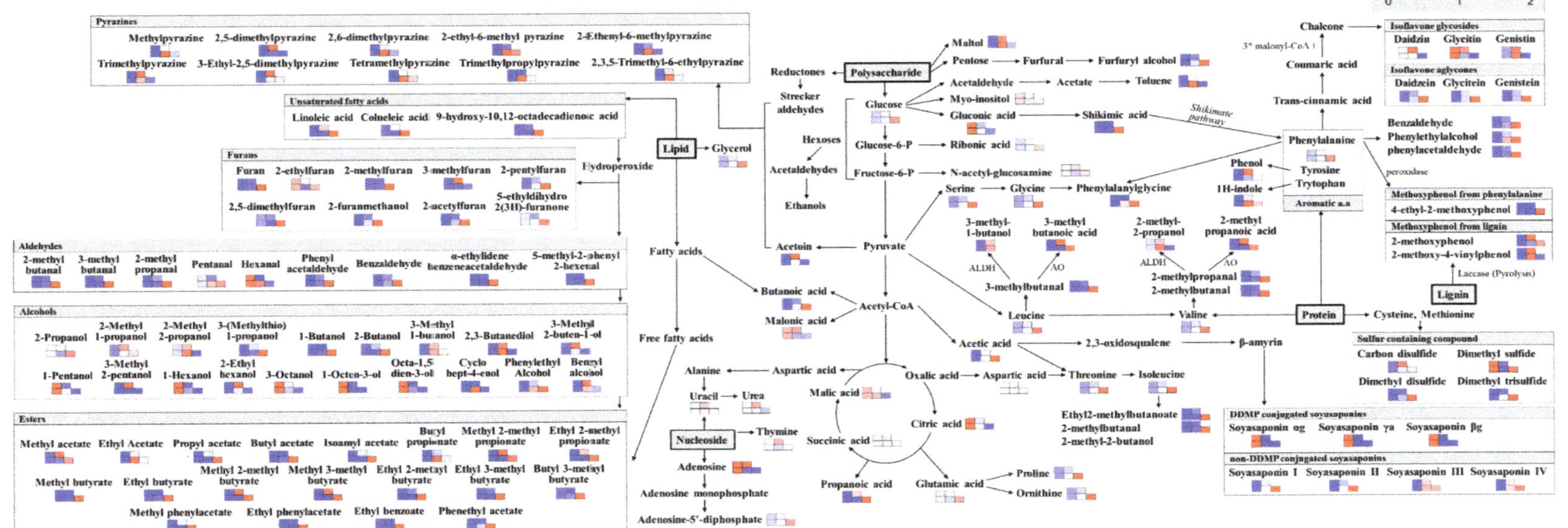

Figure 3. Schematics showing the relative abundance of the significantly discriminant metabolites (primary and secondary metabolites, as well as VOCs), detected comprehensively from the fermented soy foods (cheonggukjang, meju, and doenjang) and unfermented soybean. The metabolites are shown in the form of biosynthetic pathways, adapted from the Kyoto Encyclopedia of Genes and Genomes (KEGG) database, relatively higher in cheonggukjang. Nucleotide compounds were relatively more abundant in cheonggukjang.

3.2.2. Secondary Metabolite Profiling of Soybean and Its Fermented Foods

Totally, 29 significantly discriminant metabolites, including 10 isoflavones, 6 soyasaponins, 7 peptides, and 3 lipids, were selected at a VIP value > 1.0 and p-value < 0.05, using the PLS-DA model based on LC-MS datasets (Table S4). Isoflavones and soyasaponins were major metabolites contributing to the observed variance between the metabolite profiles of the different soy products (Figure 3). Notably, the relative levels of β-glucoside derivatives, including daidzin, glycitin, and genistin, were significantly lower in fermented soy products, particularly in doenjang. Glycitin contents were highest in soybean, while daidzin contents were observed more in cheonggukjang. Highest genistin contents were recorded from meju and cheonggukjang. Aglycone, such as daidzein, glycitein, and genistein, contents were recorded highest in doenjang and lowest in soybean. Of the soyasaponins detected in this experiment, all DDMP-conjugated soyasaponins, including soyasaponin αg, βa, and βg, were observed in relatively higher quantities in soybean extracts. Meanwhile, non-DDMP-conjugated soyasaponins, including I, II, III, and IV, were observed in significantly higher quantities in fermented soy products, especially the longer fermented doenjang samples, as compared to the raw and unfermented soybean samples.

3.2.3. Volatile Compounds Profiling of Soybean and Its Fermented Foods

A total of 148 volatile compounds, including 18 organic acids, 20 alcohols, 18 esters, 10 aldehydes, 14 ketones, 8 alkanes/alkenes, 11 aromatic compounds, 4 fatty esters, 11 furans, 5 phenols, 11 pyrazines, and 5 sulfur compounds, were determined as significantly discriminant among different soy products (Table S5). The analysis of flavor-related volatile compounds, using HS-SPME/GC-TOF-MS, indicated that most of the carboxylic acids, aldehydes, esters, furans, sulfur containing compounds, and methoxyphenol displayed higher relative abundance in doenjang, compared to other fermented soy products and soybean. A few of the flavor compounds, including maltol, branched-chain carboxylic acids, and methoxyphenol (from lignin and pyrazine), were relatively higher in cheonggukjang, compared to meju, doenjang, and soybean.

The VOCs profiles detected for doenjang contained higher relative abundance of carboxylic acids, aldehydes, most esters, furans, sulfur containing compounds, and methoxyphenol from phenylalanine. However, the relative levels of maltol and branched-chain carboxylic acids were higher in cheonggukjang. Pyrazine compounds mainly showed the highest content in meju or cheonggukjang, and compounds that were higher in meju also tended to deplete in doenjang, with longer durations of fermentation.

3.3. Correlation Assay between Total Phenolics, Flavonoids, Antioxidant Properties, and Metabolites

To compare the flavonoid, phenolic contents, and antioxidant activities of soybean and fermented soy products, ABTS, FRAP, total phenolic content (TPC), and total flavonoid content (TFC) were measured (Figure 4). According to the results of the ABTS and FRAP assay, doenjang showed significantly higher antioxidant activity, as compared to cheonggukjang, meju, and unfermented soybean samples. The TPC value was recorded the highest for doenjang among all samples. Notably, the TPC values for fermented (cheonggukjang and meju) and unfermented soybean varied marginally. Unlike other bioactivity assay results, the TFC as showed insignificantly different among fermented soy products and unfermented soybean.

Correlation analysis between the metabolites profiling results and bioactivity assays, such as ABTS, FRAP, TFC, and TPC, were conducted and illustrated in Figure 5. ABTS, FRAP, and TPC showed positive correlations with isoflavone aglycones, such as daidzein, glycitein, and genistein, and negative correlations with β-glucosides, such as daidzin, glycitin, and genistin. In case of soyasaponin, non-DDMP soyasaponins, such as I, II, III, and IV, showed a positive correlation with ABTS, FRAP, and TPC. However, DDMP-conjugated soyasaponins, such as αg, βa, and βg, showed negative correlations with FRAP. Amino acids and dipeptide showed positive correlation with ABTS, FRAP, and TPC values.

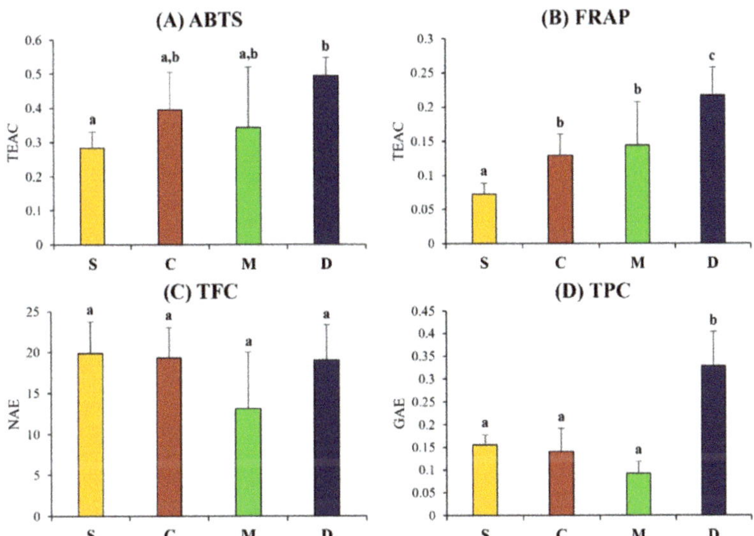

Figure 4. Total flavonoid, phenolic contents, and antioxidant activity tests. (**A**) 2,2-azino-bis-(3-ethylbenzothiazoline-6-sulfonic acid) diammonium salt (ABTS) and (**B**) ferric reducing antioxidant power (FRAP), as well as (**C**) total flavonoid contents (TFC) and (**D**) total phenolic contents (TFC). S, soy; C, cheonggukjang; M, meju; D, doenjang. Values are expressed as the average of three biological replicates ($n = 3$). Bar graphs denoted by the same letter were not significantly different, according to Duncan's multiple range test (p-value < 0.05).

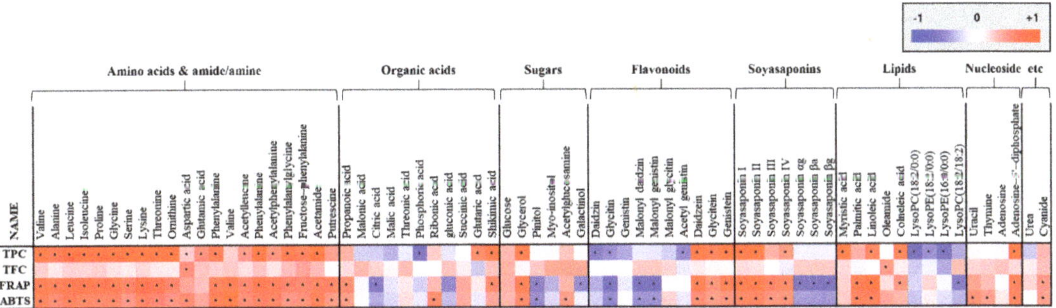

Figure 5. Heat map representation of correlation analysis between relative abundance of significantly discriminant metabolites and antioxidant activity (ABTS and FRAP), total flavonoid content (TFC), and total phenolic content (TPC). Each square indicates Pearson's correlation coefficient values (r). Red and blue represent positive ($0 < r < 1$) and negative ($-1 < r < 0$) correlation, respectively. *; p-value less than 0.05, according to Duncan's multiple range test.

4. Discussion

Soybean products are widely consumed as fermented foods with varying taste, aroma, and nutritional ingredients, owing to different fermentative bioprocesses. While comparisons between fermented soy foods, manufactured in various ways, have widely studied [9,15,16]. The comprehensive understanding how different fermentative bioprocesses influence end-product metabolomes, including the VOCs, is not completely achieved. Herein, to understand influence of fermentation methods and duration on final fermented soy foods, various fermented soy products (cheonggukjang, meju, and doenjang) with raw materials (soybean) were compared, using metabolite profiling platforms. To do so, multiple hyphenated MS-based platforms GC–TOF-MS, UHPLC-LTQ-Orbitrap-MS, and HS-SPME/GC-TOF-MS were employed. Most primary metabolite contents were significantly higher in doenjang, which can be affected by the long fermentation period (3–6 months), including the brining and aging steps, where both the substrate and microbes contributed to the metabolite pool [17]. Moreover, the fermentative microorganisms of doenjang are usually more diverse and enriched, as compared to those for cheonggukjang or meju, with axenic cultures of *Bacillus* or *Aspergillus* [18–20].

Organic acids influence the features of fermented soy products as a flavoring agent by imparting the acidity and sweet flavor. The relatively lower amount of organic acids in doenjang extracts was detected. The organic acids might be consumed as the substrates by microbes during the doenjang fermentation [21] or converted to VOCs [11] (reference [22]). Further, amino acids could be produced from a few organic acid intermediates during the glycolysis and/or TCA cycles [20]. It is known that the higher contents of amino acids are known to influence certain organoleptic traits, including sweetness, bitterness, and savory, in fermented end-products [22]. In this study, the significantly higher levels of most all amino acids was observed in doenjang extracts, as compared to meju, cheonggukjang, and raw soybean.

Considering the secondary metabolite profiles, isoflavones and soyasaponins were the major compounds that varied markedly between fermented soy products (cheonggukjang, meju, and doenjang) and unfermented soybean. Among isoflavones, in raw soybean, significantly higher amounts of isoflavone glucosides (daidzin, glycitin, and genistin) were detected. On the other hand, the isoflavone aglycones (daidzein, glycitein, and genistein) showed comparatively higher contents in doenjang. It is known that the isoflavone glucosides in soybeans are converted to isoflavone aglycones by the β-glucosidases released by the fermentative microorganisms [23]. Moreover, the relative abundance of non-DDMP-conjugated soyasaponins were increased in fermented products, according to the periods of fermentation, i.e., cheonggukjang < meju < doenjang, as described in a previous study [24]. Expectedly, the DDMP-conjugated soyasaponin contents were relatively higher in unfermented soybean. The degradation of DDMP-conjugated soyasaponin could be affected by the unstable characteristic of the DDMP conjugated form and food fermentation [25,26].

Among VOCs, relatively higher amounts of maltol, contributing the sweet flavor [27], were detected in fermented soy products than in raw materials, with a relatively higher abundance in cheonggukjang than doenjang [11]. Generally, maltol is produced by the Maillard reaction of sugars under high temperature and pressure [28]. Hence, we conjecture that the heat applied with rice straw during the soybean pre-processing, prior to cheonggukjang fermentation, might affect higher maltol levels. Additionally, a significantly higher abundances of 3-methylbutanoic acid and 3-methylbutanol were observed in cheonggukjang, which could be produced from 3-methylbutanal by oxidoreductases, according to previous study [29]. Moreover, 2-methoxyphenol and 2-methoxy-4-vinylphenol contents were relatively higher in cheonggukjang. Methoxyphenol is a volatile compound that produces a unique smoky, phenolic odor in fermented foods [30]. These compounds are produced by pyrolysis, followed by laccase-mediated transformation of lignin by *Aspergillus* [31]. Furthermore, in cheonggukjang and meju, a relatively higher abundance of pyrazines was observed in doenjang, than was in previous studies [32,33], which could be influenced by *B. subtilis* [34], originated from rice straws.

The discrepancy of metabolite contents in different fermented soy foods be caused by the variations in substrate pre-process and fermentation duration. As a result, the total flavonoid and phenolic contents, as well as the antioxidant activities, of various fermented soy products were varied. Doenjang showed significantly higher ABTS, FRAP, and TPC value, as compared to cheonggukjang, meju, and unfermented soybean samples. Similarly, a previous study represented the formation of phenolic compounds, including daidzein, glycitein, and genistein, during fermentation and maturation, which partly affected the high antioxidant activities of fermented products [35].

Further, the statistical correlation analysis was performed to explore particular metabolites, which contribute to bioactivities. Most amino acids, aglycone derivatives of flavonoids, non-DDMP soyasaponins, and fatty acids showed a strong positive correlation with ABTS,

FRAP, and TPC. This pattern was in accordance with the previous studies, which have reported on the high antioxidant activities of isoflavone aglycones and non-DDMP soy-asaponins in doenjang [16]. Although, it was reported that amino acids have weak antioxidant effects [36], the higher abundance of amino acids in doenjang might enhance its antioxidant activities.

5. Conclusions

In order to characterize the food quality, a comprehensive metabolite profiling of different soy fermented products and raw soybeans was carried out, using untargeted metabolite profiling approaches. As a result, in doenjang, the distinctly higher abundance of most organic acids, isoflavone aglycones and non-DDMP soyasaponins, amino acids, and fatty acids might influence its bioactivities. Moreover, the higher VOCs levels in doenjang significantly contributes to the rich taste and aroma of fermented final product. In meju, the fermentative intermediate of doenjang, the relative levels of some pyrazines and alcohols were observed as higher in meju than other soy products, due to the different fermentative processes, such as weaving with rice straws and drying. For cheonggukjang, certain aroma compounds relevant with unique odor, such as pyrazines and methoxyphenols, were more abundant than other soy products, which could be affected by the microflora from rice straws and substrate pre-processing. Based on current results, the observed variations in the metabolomes and volatolomes of soy products might be influenced by varying processing steps, microbial successions, and fermentation duration. This study may help in experimental design, process optimization, and quality control of the soy fermented foods for future research.

Supplementary Materials: The following are available online at https://www.mdpi.com/article/10.3390/foods10112516/s1, Table S1: Sample information of 5 soy, 5 cheonggukjang, 5 meju, and 5 doenjang. Table S2: The extraction yield of the samples. Table S3: Discriminative metabolites in soy, cheonggukjang, meju, and doenjang, derived from the PLS-DA model of the UHPLC-LTQ-Orbitrap-MS analysis. Table S4: Discriminative metabolites in soy, cheonggukjang, meju, and doenjang, derived from the PLS-DA model of the GC–TOF-MS analysis. Table S5: Discriminative metabolites in soy, cheonggukjang, meju, and doenjang, derived from the PLS-DA model of the HS-SPME/GC-TOF-MS analysis.

Author Contributions: Conceptualization, S.-H.L. (Seung-Hwa Lee), H.-J.K. and C.-H.L.; data curation, S.-H.L. (Sang-Hee Lee) and S.L.; formal analysis, S.-H.L. (Sang-Hee Lee); investigation, S.-H.L. (Sang-Hee Lee); resources, S.-H.L. (Seung-Hwa Lee) and H.-J.K.; supervision, C.-H.L.; visualization, S.-H.L. (Sang-Hee Lee); Writing—original draft, S.-H.L. (Sang-Hee Lee); Writing—review and editing, S.L., D.S. and C.-H.L. All authors have read and agreed to the published version of the manuscript.

Funding: This work was supported by Korea Institute of Planning and Evaluation for Technology in Food, Agriculture and Forestry (IPET), through Agricultural Microbiome R&D Program (The Strategic Initiative for Microbiomes in Agriculture and Food), funded by the Ministry of Agriculture, Food and Rural Affairs (MAFRA) (grant number 918011-04-3-HD020). Additionally, this study was supported by the Basic Research Lab program (grant number 2020R1A4A1018648), through the National Research Foundation grant, funded by the Ministry of Science and ICT, Republic of Korea. This research was supported by a grant from the National Agricultural Products Quality Management Service in 2019. This paper was supported by Konkuk University Researcher Fund in 2019.

References

1. Feron, G.; Bonnarme, P.; Durand, A. Prospects for the Microbial Production of Food Flavours. *Trends Food Sci. Technol.* **1996**, *7*, 285–293. [CrossRef]
2. Blandino, A.; Al-Aseeri, M.E.; Pandiella, S.S.; Cantero, D.; Webb, C. Cereal-Based Fermented Foods and Beverages. *Food Res. Int.* **2003**, *36*, 527–543. [CrossRef]

3. Egounlety, M.; Aworh, O.C. Effect of Soaking, Dehulling, Cooking and Fermentation with Rhizopus Oligosporus on the Oligosaccharides, Trypsin Inhibitor, Phytic Acid and Tannins of Soybean (Glycine Max Merr.), Cowpea (Vigna Unguiculata L. Walp) and Groundbean (Macrotyloma Geocarpa Harms). *J. Food Eng.* **2003**, *56*, 249–254. [CrossRef]

4. Singh, D.; Lee, S.; Lee, C.H. Metabolomics for Empirical Delineation of the Traditional Korean Fermented Foods and Beverages. *Trends Food Sci. Technol.* **2017**, *61*, 103–115. [CrossRef]

5. Shin, D.; Jeong, D. Korean Traditional Fermented Soybean Products: Jang. *J. Ethn. Foods* **2015**, *2*, 2–7. [CrossRef]

6. Lee, S.; Do, S.-G.; Kim, S.Y.; Kim, J.; Jin, Y.; Lee, C.H. Mass Spectrometry-Based Metabolite Profiling and Antioxidant Activity of Aloe Vera (Aloe Barbadensis Miller) in Different Growth Stages. *J. Agric. Food Chem.* **2012**, *60*, 11222–11228. [CrossRef]

7. Musteata, F.M.; Vuckovic, D. In Vivo Sampling with Solid-Phase Microextraction. In *Handbook of Solid Phase Microextraction*; Elsevier: Amsterdam, The Netherlands, 2012; pp. 399–453. [CrossRef]

8. Jeleń, H.H.; Wlazły, K.; Wasowicz, E.; Kamiński, E. Solid-Phase Microextraction for the Analysis of Some Alcohols and Esters in Beer: Comparison with Static Headspace Method. *J. Agric. Food Chem.* **1998**, *46*, 1469–1473. [CrossRef]

9. Kwon, Y.S.; Lee, S.; Lee, S.H.; Kim, H.J.; Lee, C.H. Comparative Evaluation of Six Traditional Fermented Soybean Products in East Asia: A Metabolomics Approach. *Metabolites* **2019**, *9*, 183. [CrossRef] [PubMed]

10. Lee, S.; Oh, D.-G.; Singh, D.; Lee, J.S.; Lee, S.; Lee, C.H. Exploring the Metabolomic Diversity of Plant Species across Spatial (Leaf and Stem) Components and Phylogenic Groups. *BMC Plant Biol.* **2020**, *20*, 39. [CrossRef] [PubMed]

11. Jo, Y.-J.; Cho, I.H.; Song, C.K.; Shin, H.W.; Kim, Y.-S. Comparison of Fermented Soybean Paste (Doenjang) Prepared by Different Methods Based on Profiling of Volatile Compounds. *J. Food Sci.* **2011**, *76*, C368–C379. [CrossRef]

12. Jung, E.S.; Park, H.M.; Lee, K.-E.; Shin, J.-H.; Mun, S.; Kim, J.K.; Lee, S.J.; Liu, K.-H.; Hwang, J.-K.; Lee, C.H. A Metabolomics Approach Shows That Catechin-Enriched Green Tea Attenuates Ultraviolet B-Induced Skin Metabolite Alterations in Mice. *Metabolomics* **2014**, *11*, 861–871. [CrossRef]

13. Kwon, M.C.; Kim, Y.X.; Lee, S.; Jung, E.S.; Singh, D.; Sung, J.; Lee, C.H. Comparative Metabolomics Unravel the Effect of Magnesium Oversupply on Tomato Fruit Quality and Associated Plant Metabolism. *Metabolites* **2019**, *9*, 231. [CrossRef]

14. Jung, E.S.; Lee, S.; Lim, S.H.; Ha, S.H.; Liu, K.H.; Lee, C.H. Metabolite Profiling of the Short-Term Responses of Rice Leaves (Oryza Sativa Cv. Ilmi) Cultivated under Different LED Lights and Its Correlations with Antioxidant Activities. *Plant Sci.* **2013**, *210*, 61–69. [CrossRef] [PubMed]

15. Chai, C.; Ju, H.K.; Kim, S.C.; Park, J.H.; Lim, J.; Kwon, S.W.; Lee, J. Determination of Bioactive Compounds in Fermented Soybean Products Using GC/MS and Further Investigation of Correlation of Their Bioactivities. *J. Chromatogr. B* **2012**, *880*, 42–49. [CrossRef] [PubMed]

16. Kim, T.-K.; Lee, J.-I.; Kim, J.-H.; Mah, J.-H.; Hwang, H.-J.; Kim, Y.-W. Comparison of ELISA and HPLC Methods for the Determination of Biogenic Amines in Commercial Doenjang and Gochujang. *Food Sci. Biotechnol.* **2011**, *20*, 1747–1750. [CrossRef]

17. Lee, S.Y.; Lee, S.; Lee, S.; Oh, J.Y.; Jeon, E.J.; Ryu, H.S.; Lee, C.H. Primary and Secondary Metabolite Profiling of Doenjang, a Fermented Soybean Paste during Industrial Processing. *Food Chem.* **2014**, *165*, 157–166. [CrossRef] [PubMed]

18. Kim, M.J.; Kwak, H.S.; Jung, H.Y.; Kim, S.S. Microbial Communities Related to Sensory Attributes in Korean Fermented Soy Bean Paste (Doenjang). *Food Res. Int.* **2016**, *89*, 724–732. [CrossRef]

19. Jung, J.Y.; Lee, S.H.; Jeon, C.O. Microbial Community Dynamics during Fermentation of Doenjang-Meju, Traditional Korean Fermented Soybean. *Int. J. Food Microbiol.* **2014**, *185*, 112–120. [CrossRef]

20. Nam, Y.-D.; Yi, S.-H.; Lim, S.-I. Bacterial Diversity of Cheonggukjang, a Traditional Korean Fermented Food, Analyzed by Barcoded Pyrosequencing. *Food Control* **2012**, *28*, 135–142. [CrossRef]

21. Kim, E.J.; Hong, J.Y.; Shin, S.R.; Heo, H.J.; Moon, Y.S.; Park, S.H.; Kim, K.S.; Yoon, K.Y. Taste composition and biological activities of cheonggukjang containing Rubus coreanum. *Food Sci. Biotechnol.* **2008**, *17*, 687–691.

22. Anson, L. The Bitter-Sweet Taste of Amino Acids. *Nature* **2002**, *416*, 136. [CrossRef]

23. Gottschalk, G. Regulation of Bacterial Metabolism. In *Bacterial Metabolism*; Springer: New York, NY, USA, 1986; pp. 178–207. [CrossRef]

24. Kum, S.-J.; Yang, S.-O.; Lee, S.M.; Chang, P.-S.; Choi, Y.H.; Lee, J.J.; Hurh, B.S.; Kim, Y.-S. Effects of Aspergillus Species Inoculation and Their Enzymatic Activities on the Formation of Volatile Components in Fermented Soybean Paste (Doenjang). *J. Agric. Food Chem.* **2015**, *63*, 1401–1418. [CrossRef]

25. Kim, J.; Choi, J.N.; Kang, D.; Son, G.H.; Kim, Y.-S.; Choi, H.-K.; Kwon, D.Y.; Lee, C.H. Correlation between Antioxidative Activities and Metabolite Changes during Cheonggukjang Fermentation. *Biosci. Biotechnol. Biochem.* **2011**, *75*, 1102282394. [CrossRef]

26. Yoshiki, Y.; Okubo, K. Active Oxygen Scavenging Activity of DDMP (2,3-Dihydro-2,5-Dihydroxy-6-Methyl-4H-Pyran-4-One) Saponin in Soybean Seed. *Biosci. Biotechnol. Biochem.* **2014**, *59*, 1556–1557. [CrossRef]

27. Fors, S. Sensory Properties of Volatile Maillard Reaction Products and Related Compounds. *ACS Symp. Ser.* **1983**, *215*, 185–286.

28. Yaylayan, V.A.; Mandeville, S. Stereochemical Control of Maltol Formation in Maillard Reaction. *J. Agric. Food Chem.* **2002**, *42*, 771–775. [CrossRef]

29. Smit, B.A.; Engels, W.J.M.; Smit, G. Branched Chain Aldehydes: Production and Breakdown Pathways and Relevance for Flavour in Foods. *Appl. Microbiol. Biotechnol.* **2009**, *81*, 987–999. [CrossRef]

30. Maga, J.A.; Katz, I. Simple Phenol and Phenolic Compounds in Food Flavor. *Crit. Rev. Food Sci. Nutr.* **2009**, *10*, 323–372. [CrossRef]

31. Pollegioni, L.; Tonin, F.; Rosini, E. Lignin-Degrading Enzymes. *FEBS J.* **2015**, *282*, 1190–1213. [CrossRef] [PubMed]

32. Rizzi, G.P. The Biogenesis of Food-related Pyrazines. *Food Rev. Int.* **2009**, *4*, 375–400. [CrossRef]

33. Sugawara, E.; Ito, T.; Yonekura, Y.; Sakurai, Y.; Odagiri, S. Effect of Amino Acids on Microbiological Pyrazine Formation by B. Natto in a Chemically Defined Liquid Medium. *Nippon. Shokuhin Kogyo Gakkaishi* **1990**, *37*, 520–523. [CrossRef]
34. David Nancy Gary Kippinga, J.; Amesa Owens, J.M. Formation of Volatile Compounds During Fermentation of Soya Beans Bacillus Subtilis. *J. Sci. Food Agric.* **1997**, *74*, 132–140. [CrossRef]
35. Ju, O.H.; Kim, C.-S. Antioxidant and Nitrite Scavenging Ability of Fermented Soybean Foods (Chungkukjang, Doenjang). *J. Korean Soc. Food Sci. Nutr.* **2007**, *36*, 1503–1510.
36. Lee, D.E.; Shin, G.R.; Lee, S.; Jang, E.S.; Shin, H.W.; Moon, B.S.; Lee, C.H. Metabolomics Reveal That Amino Acids Are the Main Contributors to Antioxidant Activity in Wheat and Rice Gochujangs (Korean Fermented Red Pepper Paste). *Food Res. Int.* **2016**, *87*, 10–17. [CrossRef]

Article

Metabolite–Flavor Profile, Phenolic Content, and Antioxidant Activity Changes in Sacha Inchi (*Plukenetia volubilis* L.) Seeds during Germination

Kannika Keawkim [1,2], **Yaowapa Lorjaroenphon** [1] , **Kanithaporn Vangnai** [1] **and Kriskamol Na Jom** [1,*]

[1] Department of Food Science and Technology, Faculty of Agro-Industry, Kasetsart University, Bangkok 10900, Thailand; k.kannika2252@gmail.com (K.K.); yaowapa.l@ku.th (Y.L.); kanithaporn.v@ku.th (K.V.)
[2] Faculty of Science and Technology, Huachiew Chalermprakiet University, Bang Chalong 10540, Thailand
* Correspondence: kriskamol.n@ku.ac.th

Citation: Keawkim, K.; Lorjaroenphon, Y.; Vangnai, K.; Jom, K.N. Metabolite–Flavor Profile, Phenolic Content, and Antioxidant Activity Changes in Sacha Inchi (*Plukenetia volubilis* L.) Seeds during Germination. *Foods* **2021**, *10*, 2476. https://doi.org/10.3390/foods10102476

Academic Editors: Yelko Rodríguez-Carrasco and Bojan Šarkanj

Received: 7 September 2021
Accepted: 13 October 2021
Published: 16 October 2021

Publisher's Note: MDPI stays neutral with regard to jurisdictional claims in published maps and institutional affiliations.

Abstract: Sacha inchi seeds are abundant in nutrients such as linolenic acids and amino acids. Germination can further enhance their nutritional and medicinal value; however, germination time is positively correlated with off-flavor in germinated seeds. This study investigated the changes in the metabolite and flavor profiles and evaluated the nutritional quality of sacha inchi seeds 8 days after germination (DAG). We also determined their phenolic content and antioxidant activity. We used gas chromatography equipped with a flame ionization detector (GC-FID) and gas chromatography–mass spectrometry (GC-MS) and identified 63 metabolites, including 18 fatty acid methyl esters (FAMEs). FAMEs had the highest concentration in ungerminated seeds, especially palmitic, stearic, linoleic, linolenic, and oleic acids. Amino acids, total phenolic compounds (TPCs), and antioxidant activity associated with health benefits increased with germination time. At the final germination stage, oxidation products were observed, which are associated with green, beany, and grassy odors and rancid and off-flavors. Germination is a valuable processing step to enhance the nutritional quality of sacha inchi seeds. These 6DAG or 8DAG seeds may be an alternative source of high-value-added compounds used in plant-protein-based products and isolated protein.

Keywords: antioxidant capacity; flavoromics; germination; metabolomics; *Plukenetia volubilis* L.

1. Introduction

Sacha inchi (*Plukenetia volubilis* Linneo), from the Euphorbiaceae family, also known as inca peanut, wild peanut, or sacha peanut, is generally cultivated in many regions of the Peruvian Amazon [1]. Sacha inchi has significant potential economic value in cosmetic, pharmaceutical, and food industries, and has recently been introduced as an alternative crop in Thailand. In 2020, the global Sacha Inchi market volume was valued at 92 Million USD in 2020. However, it is expected to receive a higher market share with a CAGR of 4.7% to reach 125.8 million USD during 2021–2027 [2]. The increasing demand for vegan, organic products, protein-rich food products, and supplements, as well as its demand in the food and beverages industry and the cosmetics sector, is expected to push the sacha inchi industry growth [3]. At present, several commercial products have increased availability in particular oil, roasted seeds, and protein powder or flour.

Sacha inchi seeds are nutritionally superior as a good source of oil (~41%), protein (~27%), and bioactive compounds [4], such as polyphenolic compounds, phytosterols (stigmasterol and campesterol), and tocopherols. The oil is rich in unsaturated fatty acids (~93% of total lipids). Essential fatty acids in the seeds comprise 33% α-linoleic acid and ~51% α-linolenic acid [5]. The seeds contain high amounts of essential amino acids, such as isoleucine, valine, and tryptophan [6]. However, they also contain antinutritional components, such as saponins, tannins, and trypsin inhibitors [7,8], and are usually consumed roasted because of the astringency due to antinutritional components. The germination conditions of sacha inchi could reduce tannins and other antinutritional components and enhance the seeds' nutritional quality [9].

Germination has been studied in many seed plants, such as soybean [10], barley [11], mung bean [12], sunflower [13], and mangosteen [14]. A germinating seed depends on reserve carbohydrates, proteins, and lipids synthesized and stored during seed growth. These compounds are synthesized during germination via glycolysis, the tricarboxylic acid (TCA) cycle, and amino acid metabolic pathways. Germination begins with water imbibition, food reserve mobilization, protein synthesis, and consequent radicle protrusion—the main food reserve for seedling development. The food reserves may be stored mainly in the endosperm (many monocotyledons, cereal grains, and castor) or cotyledons (many dicotyledons, such as peas and beans). The macronutrients (proteins, lipids, and carbohydrates) are enzymatically (amylases, proteases) hydrolyzed into simple units of food comprising sugars and amino acids. These dissolve in water and pass toward the growing epicotyl, hypocotyl, radicle, and plumule through the cotyledons [15]. Triacylglycerol (storage lipids) metabolism during germination has also been reported in sacha inchi. During germination, triacylglycerols are hydrolyzed into fatty acids by lipase via lipid metabolism. The released fatty acids undergo β-oxidation to generate adenosine triphosphate (ATP) as the major energy supply within the seeds [16].

However, germination time is positively correlated to off-flavor in germinated seeds. The activity of lipoxygenases (LOXs) on storage lipids in seeds can produce oxidation products, such as aldehydes and alcohols, leading to unpleasant beany, green, or rancid flavors [17]. The metabolites and flavor compounds that change during germination can be used as biomarkers according to the germination stages in sacha inchi.

Metabolomics is the scientific study of multiple small molecules in cells, tissues, plants, or foods. The metabolite profile contains valuable data for breeding-driven metabolic engineering of nutritionally important metabolites in plants. Metabolomics has been applied in plant and food science to study the effect of germination on mung bean [12] and barley [11] and the effect of processing and storage on Nam Dok Mai mango wine fermentation [18]. The metabolomics approach can be targeted or nontargeted. Nontargeted metabolomics can be used to simultaneously analyze hundreds or thousands of known and unknown metabolites. In addition, flavor metabolomics or flavoromics can be applied in various food products, focusing on the changes in the targeted and nontargeted flavor compounds.

There is still a lack of research information on the germination of sacha inchi seed, as well as the metabolism of triacylglycerols during seed germination of sacha inchi [19]. Only some limited information of the changes of fatty acids during germination of sacha inchi was investigated without any report on other nutrients.

Considering the insufficient information about the changes of metabolites, flavor compounds, as well as total phenolic compounds and antioxidant activity during germination of sacha inchi, therefore, this study investigated the different metabolites and flavor components formed during germination in sacha inchi seeds using coupled metabolomics–flavoromics analysis. We also investigated the phenolic content and antioxidant activity at different germination times. In addition, we identified important biomarkers during germination in order to determine the best nutritional values and acceptability of foods and ingredients made from sachi inchi seeds in the future.

2. Materials and Methods

2.1. Chemicals

All chemical solvents and reagents for extraction and gas chromatography (GC) derivatization were high-performance liquid chromatography (HPLC) and analytical grade and were purchased from RCI Labscan Ltd. (Pathumwan, Bangkok, Thailand) and Sigma-Aldrich (St. Louis, MO, USA), respectively. All internal standards, including tetracosane (IS I), 5α-cholestane-3β-ol (IS II), phenyl-β-D-glucopyranoside (IS III), ρ-chloro-L-phenylalanine (IS IV), and dodecanoic acid ethyl ester, were standard grade and were purchased from Sigma-Aldrich. In addition, all reference standards for metabolomics and flavoromics, a C6–C30 n-alkane mixture used to measure linear retention indices, and 2,2-diphenyl-1-picrylhydrazyl (DPPH), 2,2-azobis(3-ethylbenzothialzoline-6-sulfonic acid) (ABTS), and Folin–Ciocalteu reagent were purchased from Sigma-Aldrich.

2.2. Seed Sample Preparation

Raw sacha inchi seeds were purchased from Kaeng Khro City, Chaiyaphum Province, Thailand in 2020. Healthy intact seeds were disinfected by soaking them in 1% (active ingredient) hypochlorite bleach solution for 15 min, followed by washing thrice with sterilized distilled water. Next, the seeds were soaked in ultrapure water at 30 °C for 12 h and then placed between wet sterilized tissue papers in a sterilized container and watered regularly every 12 h. The container was kept in an incubator at 30 °C and 80% humidity [19].

Germination was defined by the appearance of an emerging radicle. The samples were kept for 2, 4, 6, and 8 days (Figure 1). The germinated samples were peeled and freeze-dried for 24 h using a Gamma 2–16 LSC freeze-drying machine (Martin Christ, Osterode am Harz, Germany), ground into a fine powder using a blender, and stored in an aluminum bag at −18 °C until analysis.

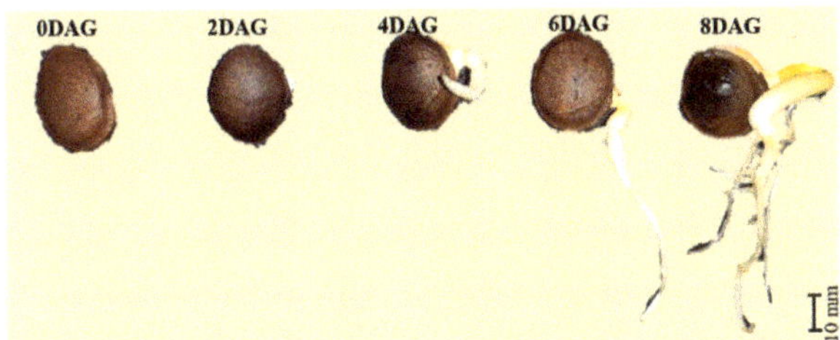

Figure 1. Germinated sacha inchi seeds at different germination days. DAG, days after germination.

2.3. Nontargeted Metabolomics

Solid-phase extraction (SPE) and fractionation of samples were performed [12]. Briefly, lipids and polar compounds (sugars, sugar alcohols, acids, amino acids, organic acids, and amines) were extracted from the samples (100 mg) using SPE many times. Dichloromethane was added to elute the lipid fraction, and the polar fraction in the defatted sample was eluted using mixed solvents (80:20 *v/v* of methanol and deionized water [DI]).

After adding 100 μL each of IS I and IS II solution to 500 μL of the lipid fraction, lipid transesterification was carried out in methanol, and consequently, lipids were separated into fractions using an SPE C18-LP cartridge (Vertical Chromatography Co., Ltd., Nonthaburi, Thailand). Fraction 1 contained fatty acid methyl esters (FAMEs) and hydrocarbons, and fraction 2 contained polar lipids (free fatty acids [FFAs], fatty alcohols, and sterols). All eluents were dried using a vacuum evaporator. The dried fraction 2 was silylated with N-methyl-N-(trimethylsilyl)trifluoroacetamide (MSTFA).

Subsequently, 250 μL each of IS III and IS IV solution was added to the polar fraction to obtain fraction 3, 1 mL of which was dried and then silylated with N-trimethylsilylimidazole (TMSIM). To obtain fraction 4, 1 mL of the polar extract was evaporated to dryness and 100 μL of the silylating agent MSTFA was added to the oximated sample and silylation performed at 70 °C for 15 min.

To obtain acids, 500 μL of hexane and 300 μL of DI water were added to the sample. After vortex mixing, the oximated sugar in the upper phase was removed and the lower phase (amino acids and free organic acids) was collected and evaporated until dryness and then silylated with MSTFA.

Finally, 1 μL of all four fractions was separately analyzed by gas chromatography equipped with a flame ionization detector (GC-FID; Hewlett Packard, Palo Alto, CA, USA) to characterize them. A DB-1 capillary column (60 m × 0.32 mm × 0.25 µm film thickness) with a 100% dimethylpolysiloxane stationary phase (Agilent Technologies, Santa Clara, CA, USA) was used to separate the components in the sample. Helium was used as the carrier gas at a constant flow rate of 1.8 mL/min. Splitless injection was performed at 280 °C. The oven temperature program started at 100 °C, then ramped up to 320 °C at 4 °C/min, and was held at 320 °C for 25 min. The detector temperature was set to 320 °C.

2.4. Nontargeted Flavoromics

Flavor compounds in sacha inchi seeds were investigated and measured using 6890N GC coupled with a time-of-flight mass spectrometer (Leco Corp., St. Joseph, MI, USA). Dodecanoic acid ethyl ester was used as an internal standard, and a group of n-alkanes (C6–C30) were used as standards to determine the retention index. Autosampler headspace solid-phase microextraction (HS-SPME) was performed to extract all volatile compounds from the sample [20].

To extract flavor compounds, ~3 g of the sample was put in a 20 mL headspace vial and 1 µL of dodecanoic acid ethyl ester (1 µg/mL in methanol) and 1 g of NaCl were added before sealing the vial cap. The sample was automatically equilibrated at 70 °C for 10 min. The volatile compounds were adsorbed on 50/30 µm divinylben-zene/carboxen/polydimethylsiloxane (DVB/CAR/PDMS) fiber at 70 °C for 30 min. Then, the fiber was desorbed into the GC injection port at 250 °C for 5 min in splitless mode.

GC 6890N coupled with a time-of-flight mass spectrometer (Leco Corp.) was used to measure the volatile compounds in the sample. A Stabilwax fused silica column (30 m × 0.25 mm × 0.25 µm film thickness) with a cross-bond polyethylene glycol station-ary phase was used to separate volatile compounds. The temperature program was started at 40 °C for 5 min, ramped up to 225 °C at 4 °C/min, and then held for 15 min. For MS, the ion source temperature was set to 200°C and the transfer line temperature to 225 °C. Mass spectra were investigated at −70 eV in the m/z range of 35–500 atomic mass units (amu). Identification and relative concentration of flavor compounds were performed using the ChromaTOF-GC software v4.50.8.0.

2.5. Determination of Total Phenolic Compounds

In the extraction method of Atala et al. (2009) with slight modifications [21], 2.5 g of the sample were extracted with 25 mL of 75:25 v/v acetone and DI water solution and shaken in a water bath at room temperature for 90 min. Then, the extract was centrifuged at $2000 \times g$ for 15 min, and the supernatant was collected and kept in a brown vial at −20 °C to evaluate the total phenolic compounds (TPCs) and antioxidant activity.

The TPCs were determined using the protocol of Chirinos et al. (2016) with slight modifications [22]. Briefly, 500 µL of the sample extract was mixed with 1250 µL of Na_2CO_3 (7.5% w/v) and 250 µL of Folin–Ciocalteu reagent and reacted for 30 min at room temperature. Next, absorbance was measured with a spectrophotometer at 755 nm wavelength against a blank in 500 µL of DI water, 1250 µL of Na_2CO_3, and 250 µL of Folin–Ciocalteu reagent. The results are expressed as gallic acid equivalents (mg GAE/g sample).

2.6. Determination of Antioxidant Activity

DPPH and ABTS assays were performed to evaluate the antioxidant activity using a spectrophotometer. The 0.1 mg/mL (75:25 v/v acetone and DI water solution) of extracted solution was used to determine antioxidant activity. Ascorbic acid was used as positive control. The antioxidant activity was expressed in terms of radical scavenging activity by the following equation:

$$\text{Radical scavenging activity (\%)} = [(A_0 - A_s)/A_s] \times 100$$

where, A_0 is absorbance of blank, and A_s is absorbance of sample extract.

In DPPH assay, as described by Brand-Williams et al. (1995) [23] with modifications, 200 µL of the sample extract was reacted with 2.8 mL of DPPH (0.1 mM in methanol) solution. After incubation for 30 min in the dark, absorbance was measured at 517 nm wavelength against a blank (200 μL of methanol in 2.8 mL DPPH solution).

In the ABTS assay, as described by Re et al. (1999) [24] with modifications, ABTS·+ solution was prepared by mixing ABTS solution (7 mmol/L) with $K_2S_2O_8$ solution (2.45 mmol/L) in a 1:1 volume ratio and stored for 12–16 h in the dark. ABTS·+ solution was diluted with ethanol in a 1:50 volume ratio to an absorbance of ~0.7 ± 0.02 at 734 nm wavelength. Then, 1 mL of ABTS·+ solution was added to 10 μL of the sample extract and reacted for 15 min in the dark. Absorbance was measured at 734 nm wavelength against a blank (10 μL of DI water in 1 mL ABTS·+ solution).

2.7. Statistical Analysis

The collected data were measured at least twice. The identical peak areas of metabolites were integrated using the HP ChemStation A.06.03 program (Hewlett Packard) and identified using the reference standards comparison method. The relative quantification was done by normalizing the corresponding peak areas to the peak area of the internal standard of each fraction and dividing the value by the weight of the extracted sample. The mass spectra of detected compounds were identified by matching mass spectra with NIST mass spectral database version 2.0 (National Institute of Standards and Technology, Gaithersburg, MD, USA). The retention time index (RI) of each volatile compound was calculated using the retention time of the n-alkanes series (C6–C30) as a reference. The relative concentration of identified compounds was quantified using the internal standard method. Dynamic changes in all metabolites and flavor compounds during germination were explained by principal component analysis (PCA), heat plots, and Spearman's rank correlation. PCA was performed using XLSTAT base software. Correlation network analysis using nonparametric Spearman's rank correlation of all metabolites and flavor compounds was performed at a significance level of $p \leq 0.05$ in the correlation/association test mode in XLSTAT.

3. Results and Discussion

3.1. Integrated Metabolomics and Flavoromics of Germinated Sacha Inchi Seeds

The germination percentage of sacha inchi seeds was ≥80%. The seeds began germinating after 2 days and reached the maximum germination percentage at 8 days. Metabolomics identified 63 (~65%) of a total of 95 metabolites at 0DAG–8DAG, similar to comparable metabolite profiling studies on mung bean [12] and soybean [25]. The 63 metabolites were categorized into 4 groups: 18 FAMEs (16 FAMEs and 2 hydrocarbons), 17 polar lipids (10 FFAs, 2 fatty alcohols, and 5 phytosterols), 6 sugars (3 organic sugars and 3 sugar alcohols), and 22 acids (17 amino and 5 organic acids). Flavoromics identified 35 (23%) of a total of 150 flavor compounds, similar to germinated brown rice (41 volatile compounds identified at 0DAG–5DAG [17]). The 35 flavor compounds were categorized into 7 groups: 9 aldehydes, 12 alcohols, 2 ketones, 8 volatile acids, 2 lactones, pyridine, pyrazine, and 2-methoxy-3-(1-methylethyl).

Figure 2 shows the PCA biplot and loading of all identified metabolites and flavor compounds at different germination times. Polar and nonpolar metabolites were major contributors to the separation along principal component (PC) 1 (Figure 2A). The flavor compounds were prominently arranged at 4DAG, 6DAG, and 8DAG, explaining 75% of the variation (Figure 2B). The PCA biplots of metabolites and flavor compounds displayed four distinct clusters along with the PC (Figure 2C). The first two PCs explained 81.67% of the total variation at 0DAG–8DAG, PC1 and PC2 demonstrated eigenvalues of 60.330 and 20.518, respectively, and explained 60.94% and 20.73% of the total variation, respectively. Metabolites and flavor compounds were categorized based on the difference in the germination times into four clusters. The differences between ungerminated seeds (0DAG) and germinated seeds could be generally described on PC1. The 0DAG seeds relied on cluster 1 (17 FAMEs, 5 phytosterols, and 8 flavor compounds), 2DAG on cluster 2 (1 FAME, 5 FFAs, and 1 flavor compound), 4DAG on cluster 3 (5 FFAs, 3 amino acids, and

4 flavor compounds), and 6DAG and 8DAG on cluster 4 (all sugars, most of the acids, and 20 flavor compounds), indicating that most of the FAMEs and FFAs decrease, while most of the sugars, amino acids, aldehydes, alchohols, and volatile acids increase throughout germination in sacha inchi.

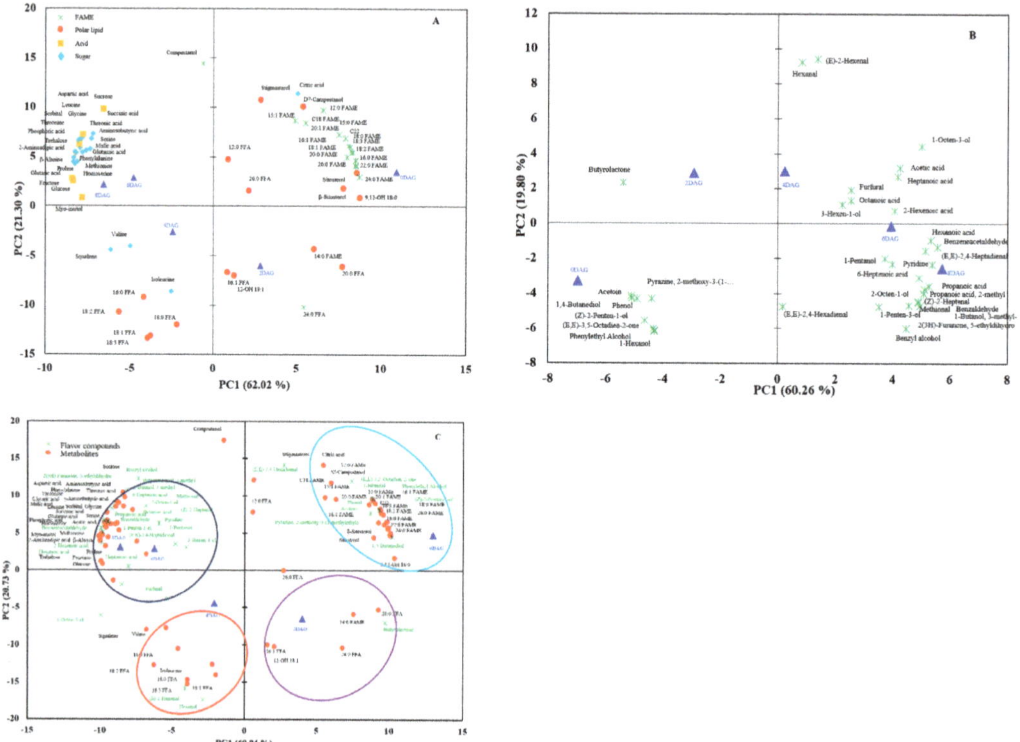

Figure 2. Biplot of principal component analysis from metabolites (**A**), flavor compounds (**B**), and metabolites and flavor compounds (**C**) in sacha inchi seeds at different germination times ($p \leq 0.05$). FAME, fatty acid methyl ester; FFA, free fatty acid; DAG, days after germination.

To confirm PCA grouping, agglomerative hierarchical clustering analysis in similarity mode was performed based on 0DAG–8DAG (Figure 3). The dendrogram confirmed the similarity among the germination stages, differentiated into four groups: group 1 included ungerminated seeds (0DAG); group 2, 2DAG seeds; group 3, 4DAG seeds; and group 4, 6DAG and 8DAG seeds. The coupled metabolomics–flavoromics of 0DAG was closer to that of 2DAG, whereas that of 4DAG was closer to that of 6DAG and 8DAG, indicating that metabolic and flavor components change more rapidly at the final compared to the early stage of germination.

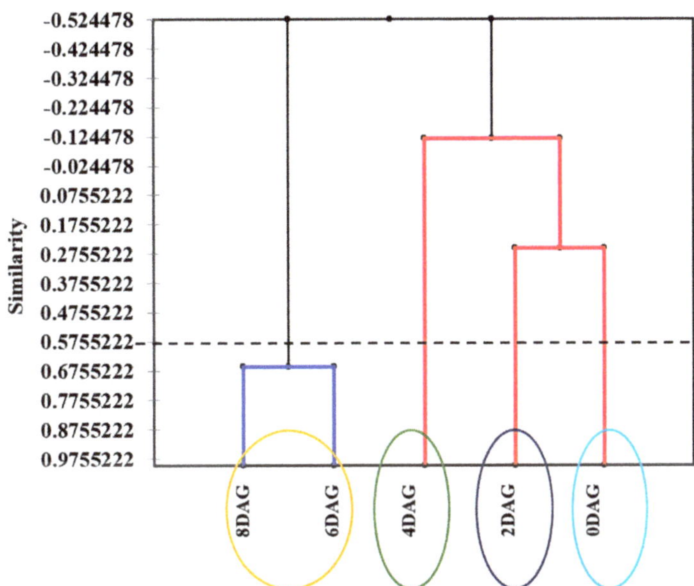

Figure 3. Dendrogram (similarity mode of agglomerative hierarchical clustering) of germinated sacha inchi seeds at different germination times ($p \leq 0.05$). DAG, days after germination.

Heat map analysis was used to represent dynamic changes in specific compounds by relative concentrations at each germination time (Figure 4). Figure 4A shows a heat map of all metabolites. The FAME contents were higher on day 0 and significantly decreased during germination. Linolenic (50%), linoleic (36%), and oleic (11%) acids were most abundant in the lipid fraction, followed by stearic (4%) and palmitic (0.25%) acids at 0DAG–8DAG. Compared to ungerminated sacha inchi seeds, the palmitic acid content decreased by ~65.6-fold, followed by stearic (13.6-fold), linoleic (12.3-fold), linolenic (11.7-fold), and oleic (4.9-fold) acids. A decrease in FAMEs has also been reported in germinated rice [26], barley [11] and mung bean [12]. During germination, triacylglycerols are hydrolyzed by lipase into FFAs and glycerol, and then hydrolyzed fatty acids are degraded into glyoxysomes by β-oxidation. Fatty acids are oxidized to carbon dioxide and water, producing energy for the germination and biosynthesis of new compounds. Therefore, palmitic, stearic, linoleic, linolenic, and oleic acids can be used as energy sources during embryo development in germinating sacha inchi. In addition, LOX is the main pathway to mobilize lipids during germination [27].

We found 10 FFAs, 2 fatty alcohols, and 5 sterols during germination. The FFA content was the highest at 2DAG, and then significantly dropped followed by a slight increase. Linolenic acid had the highest concentration at 2DAG (~27-fold), followed by oleic (13-fold), and linoleic (9-fold) acids, compared to ungerminated sacha inchi seeds. The increase in major FFAs at the beginning stage in the present study was similar to related research. Chandrasekaran et al. (2015) [19] reported that the major unsaturated fatty acids content increased at the beginning stage (3–10 days after germination), followed by a slight decrease during the germination course of sacha inchi. Under the different conditions on germination such as temperature, moisture, and germination time may influence lipid degradation rates [11] in germinated sacha inchi. At the beginning of germination, triacylglycerols are hydrolyzed by lipases into FFAs. Then, the FFAs are distributed into the mitochondrial matrix and degraded to acetyl-CoA by oxidation, which is then processed in the Krebs and glyoxylate cycles [28]. The fatty alcohol content in germinated sacha inchi seeds showed a relatively small change compared to ungerminated seeds: 9,12-OH 18:0 showed the highest concentration at 0DAG and 12-OH 18:1 at 4DAG.

FAMEs had a strong positive correlation (r > 0.7) with alcohols, ketones, butyrolactone, and pyrazine-2-methoxy-3-(1-methylethyl). However, FAMEs had a strong negative correlation (r < −0.7) with aldehydes, volatile acids, 2(3H)-furanone,5-ethyldihydro, and pyridine. In general, lipolysis and lipid oxidation produce rancidity and sourness attributed to aldehydes, alcohols, and volatile acids during germination [41]. During germination, volatile acids are formed via the lipolysis pathway. Aldehyde and alcohol formation during germination is correlated with the oxidation of oleic, linoleic, and linolenic acids as the main substrates through LOXs. These aldehydes and alcohols provide rancid and off-flavors to germinated sacha inchi seeds [42].

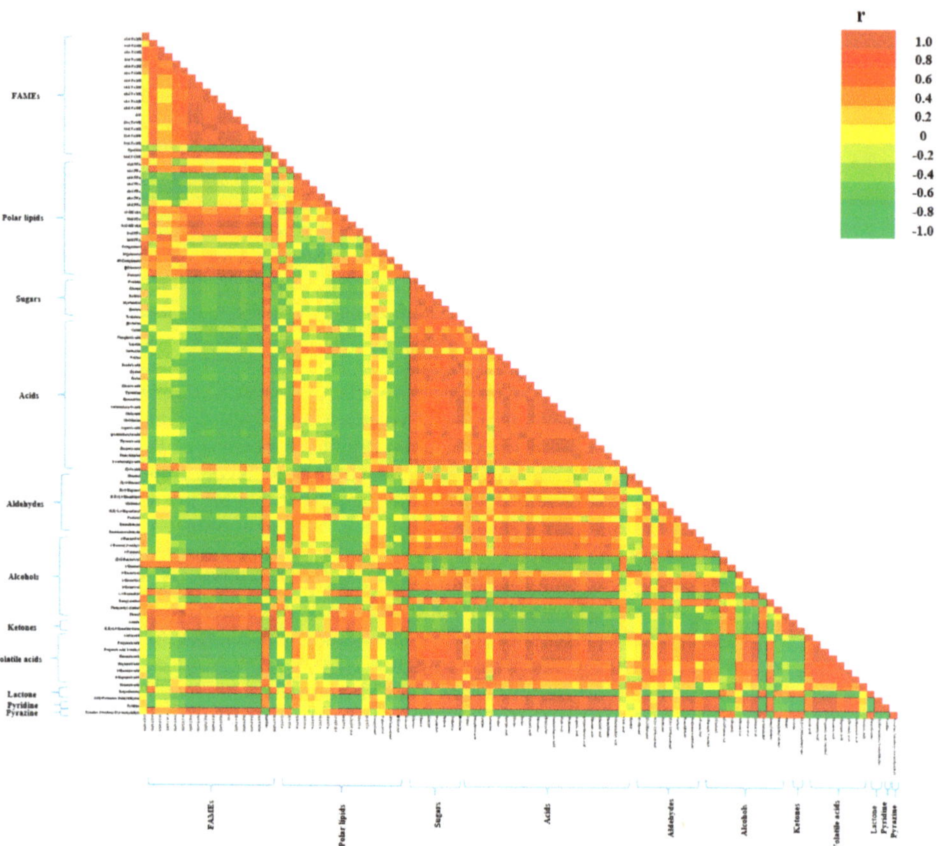

Figure 5. Lower triangular heat map represents pairwise correlation analysis between metabolites and flavor compounds during germination of sacha inchi seeds. Each square represents the Spearman's rank correlation coefficient at a significance level of $p \leq 0.05$. Orange-red, strong positive correlation (r > 0.7); green, strong negative correlation (r < −0.7). FAME, fatty acid methyl ester; FFA, free fatty acid.

Coupled metabolomics–flavoromics analysis identified 10 metabolites (palmitic acid, stearic acid, oleic acid, linoleic acid, linolenic acid, fructose, glycine, isoleucine, proline, and leucine) and 7 flavor compounds (acetoin, (E,E)-3,5-octadien-2-one, hexanal, (E)-2-hexenal,1-octen-3-ol,2-octen-1-ol, and hexanoic acid) as biomarkers of the nutritional quality of ungerminated and germinated sacha inchi seeds. Palmitic, stearic, linoleic, linolenic, and oleic acids were the most abundant fatty acids, and the highest concentration of acetoin and (E,E)-3,5-octadien-2-one was found at 0DAG. At 4DAG, excess aldehydes and alcohols (hexanal, (E)-2-hexenal, and 1-octen-3-ol) were produced, which might be considered undesirable. At 6DAG and 8DAG, the highest level of sugars and amino acids was found, which could be applied as food ingredients or plant-based protein. The highest concentration of fructose, glycine, isoleucine, proline, and leucine can be used as a biomarker to identify the germinated seed quality. In addition, the changes in metabolites and flavor compounds during germination are also significantly correlated with biological activities [43,44].

3.2. Changes in TPCs and Antioxidant Activity during Germination

Germination relies on an effective method of achieving high concentrations of bioactive compounds. Phenolic contents also increase, apart from nutrition level changes. The dynamic changes in TPCs in germinated sacha inchi seeds are shown in Figure 6. TPCs gradually increased from 0DAG to 8DAG. The impact of germination on TPCs has been studied in edible seeds, such as beans and cereal grains. Germination can continuously collect soluble phenolics in germinated edible seeds compared with ungerminated seeds [45–47]. This collection could be associated with the shikimate pathway. Phenylalanine, as an intermediate metabolite in polyphenol metabolism, increased during germination. After the seeds absorb water, hydrolytic enzymes are activated and macromolecules (biopolymers, carbohydrates, and proteins) break down into small molecules in the endosperm, followed by an increase in simple sugars and free amino acids. In addition, bound phenolic compounds conjugated with the cell wall are also released [48]. Germination can also induce oxidative tension in a plant, producing more enhanced phenolics during germination for plant antioxidant protection [49]. In sacha inchi seed germination, TPCs were higher at 8DAG than at 0DAG. In addition, the polyphenol enrichment at 8DAG may be related to the formation of off-flavor compounds in sachi inchi seeds [44].

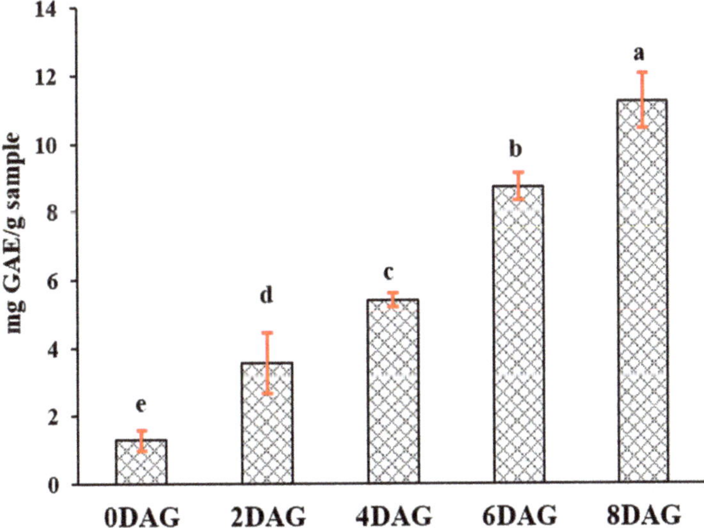

Figure 6. Total phenolic compounds (TPCs) of sacha inchi seeds at different germination times ($p \leq 0.05$). DAG, days after germination. Means followed by different letters are significantly different ($p \leq 0.05$) according to the Duncan's new Multiple Range Test.

In contrast, changes in the phenolic content were the main contributor to the antioxidant activity variation during germination [50]. The antioxidant activity of sacha inchi seeds at different germination times was evaluated by DPPH and ABTS assays (Figure 7). Germination significantly enhanced the antioxidant activity compared with nongerminated seeds. The antioxidant activity gradually increased from 0DAG to 8DAG up to

3.6- and 3-fold by DPPH and ABTS assays, respectively, and was parallel to the phenolic content, which, exhibiting a significant antioxidant potential, could be improved during germination. In addition, results showed that the highest value of the antioxidant activity on the germinated sacha inchi was stronger than the positive control of ascorbic acid at dosage of more than 10 and 20 µg/mL for DPPH and ABTS assays, respectively. Phenolics can be freed during germination, and some metabolic activities might improve due to the oxidative signaling in germinated seeds and generate energy for the developing plant. The increase in antioxidant activity might be associated with de novo synthesis and dynamic changes in phenolic compounds via phenylpropanoid biosynthesis. Germination can increase the antioxidant activity in edible seeds, such as jack bean, chick pea, soybean, kidney bean, and mung bean [48], attributed to the increase in some antioxidant components during germination, such as phenolic compounds, phytosterols, and vitamins.

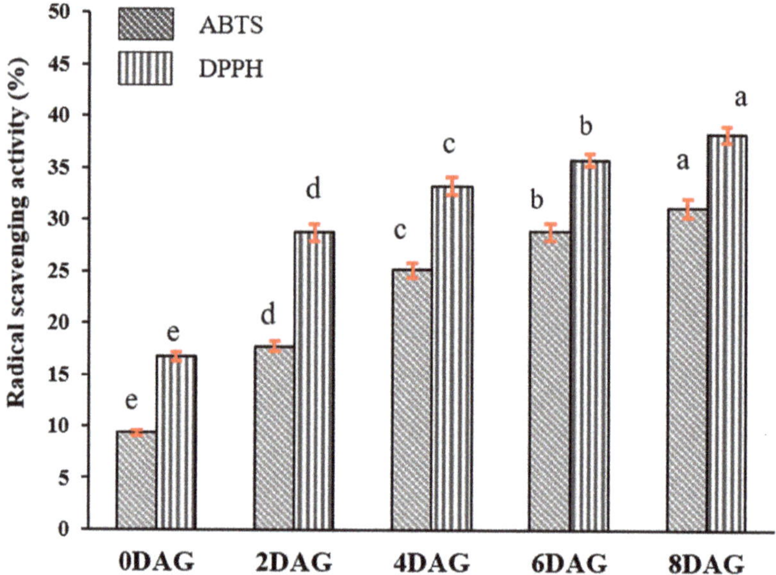

Figure 7. Antioxidant activity (ABTS and DPPH assays) of sacha inchi seeds at different germination times ($p \leq 0.05$). ABTS, 2,2-azobis(3-ethylbenzothialzoline-6-sulfonic acid; DPPH, 2,2-diphenyl-1-picrylhydrazyl; DAG, days after germination. Means followed by different letters are significantly different ($p \leq 0.05$) according to the Duncan's new Multiple Range Test.

4. Conclusions

Coupled metabolomics and flavoromics can be used to identify the main metabolites and flavor compounds that change in sacha inchi seeds during germination in order to determine the best nutritional values and guide the improvement of pharmacological products or functional foods made from sacha inchi seeds. In particular, the later stage (6–8 DAG) should be applied as novel alternative plant-based protein, functional foods, and dietary supplements, because they contain significant amounts of amino acids and higher antioxidant activity. FAMEs (palmitic, stearic, linoleic, linolenic, and oleic acids) correlate with ungerminated seeds. Sugars and amino acids (fructose, glycine, isoleucine, proline, and leucine) correlate with 8DAG seeds. Flavor compounds acetoin and (E,E)-3,5-octadien-2-one have the highest concentration in ungerminated seeds. Hexanal, (E)-2-hexenal, and 1-octen-3-ol have the highest concentration at 4DAG and 2-octen-1-ol and hexanoic acid at 8DAG. The changes in these flavor compounds are biomarkers of flavor indicators in sacha inchi seeds. Amino acids, organic acids, TPCs, and antioxidant activity are higher in germinated sacha inchi seeds than in ungerminated seeds. These findings could help in understanding nutritional changes during germination.

Author Contributions: Conceptualization, K.N.J. and K.K.; methodology, K.N.J. and K.K.; software, K.K.; validation, K.K.; formal analysis, K.K.; investigation, K.K.; resources, K.N.J., K.K., Y.L. and K.V.; data curation, K.K.; writing—original draft preparation, K.K.; writing—review and editing, K.N.J., K.K., Y.L. and K.V.; supervision, K.N.J.; funding acquisition, K.K. All authors have read and agreed to the published version of the manuscript.

Funding: This research received no external funding.

Institutional Review Board Statement: Not applicable.

Informed Consent Statement: Not applicable.

Data Availability Statement: Not applicable.

Acknowledgments: Authors are greatly thankful to the Faculty of Science and Technology of Huachiew Chalermprakiet University and Faculty of Agro-Industry, Kasetsart University for research support. Kasetsart University Research and Development Institute is acknowledged for English proofreading service and support.

Conflicts of Interest: The authors declare that they have no conflict of interest.

References

1. Hanssen, H.-P.; Schmitz-Hübsch, M. Sacha inchi (*Plukenetia volubilis* L.) nut oil and its therapeutic and nutritional uses. In *Nuts and Seeds in Health and Disease Prevention*; Preedy, V.R., Watson, R.R., Patel, V.B., Eds.; Academic Press: San Diego, CA, USA, 2011.
2. The Express Wire. Sacha Inchi Market Growth Statistics 2021, Industry Trends, Size, Share, Business Strategies, Emerging Technology, Product Portfolio, Top-Countries Data, Manufacturers Analysis, Demand Status and Forecast 2027. 2021. Available online: https://www.theexpresswire.com/pressrelease/sacha-inchi-market-growth-statistics-2021-industry-trends-size-share-business-strategies-emerging-technology-product-portfolio-top-countries-data-manufacturers-analysis-demand-status-and-forecast-2027_13740109 (accessed on 5 October 2021).
3. EMR. Global Sacha Inchi Market to Grow at a CAGR of 4% during 2021–2026, Driven by the Growing Shift towards Bodybuilding and Fitness. 2021. Available online: https://www.expertmarketresearch.com/pressrelease/global-sacha-inchi-market (accessed on 5 October 2021).
4. Chirinos, R.; Zuloeta, G.; Pedreschi, R.; Mignolet, E.; Larondelle, Y.; Campos, D. Sacha inchi (*Plukenetia volubilis*): A seed source of polyunsaturated fatty acids, tocopherols, phytosterols, phenolic compounds and antioxidant capacity. *Food Chem.* **2013**, *141*, 1732–1739. [CrossRef]
5. Gutiérrez, L. F.; Rosada, L.; Jiménez, Á. Chemical composition of Sacha Inchi (*Plukenetia volubilis* L.) seeds and characteristics of their lipid fraction. *Grasas y Aceit.* **2011**, *62*, 76–83. [CrossRef]
6. Wang, S.; Zhu, F.; Kakuda, Y. Sacha inchi (*Plukenetia volubilis* L.): Nutritional composition, biological activity, and uses. *Food Chem.* **2014**, *265*, 316–328. [CrossRef] [PubMed]
7. Rawdkuen, S.; Murdayanti, D.; Ketnawa, S.; Phongthai, S. Chemical properties and nutritional factors of pressed-cake from tea and sacha inchi seeds. *Food Biosci.* **2016**, *15*, 64–71. [CrossRef]
8. Bueno-Borges, L.B.; Sartim, M.A.; Gil, C.C.; Sampaio, S.V.; Rodrigues, P.H.V.; Regitano-d'Arce, M.A.B. Sacha inchi seeds from sub-tropical cultivation: Effects of roasting on antinutrients, antioxidant capacity and oxidative stability. *J. Food Sci. Technol.* **2018**, *55*, 4159–4166. [CrossRef] [PubMed]
9. Xu, M.; Jin, Z.; Lan, Y.; Rao, J.; Chen, B. HS-SPME-GC-MS/olfactometry combined with chemometrics to assess the impact of germination on flavor attributes of chickpea, lentil, and yellow pea flours. *Food Chem.* **2019**, *280*, 83–95. [CrossRef]
10. Cheng, L.; Gao, X.; Li, S.; Shi, M.; Javeed, H.; Jing, X.; Yang, G.; He, G. Proteomic analysis of soybean [*Glycine max* (L.) Meer.] seeds during imbibition at chilling temperature. *Mol. Breed.* **2010**, *26*, 1–17. [CrossRef]
11. Frank, T.; Scholz, B.; Peter, S.; Engel, K.-H. Metabolite profiling of barley: Influence of the malting process. *Food Chem.* **2011**, *124*, 948–957. [CrossRef]
12. Na Jom, K.N.; Frank, T.; Engel, K.H. A metabolite profiling approach to follow the sprouting process of mung beans (*Vigna radiata*). *Metabolomics* **2011**, *7*, 102–117. [CrossRef]
13. Pajak, P.; Socha, R.; Gałkowska, D.; Rożnowski, J.; Fortuna, T. Phenolic profile and antioxidant activity in selected seeds and sprouts. *Food Chem.* **2014**, *143*, 300–306. [CrossRef]
14. Mazlan, O.; Aizat, W.M.; Aziz Zuddin, N.S.; Baharum, S.N.; Noor, N.M. Metabolite profiling of mangosteen seed germination highlights metabolic changes related to carbon utilization and seed protection. *Horticulture* **2019**, *243*, 226–234. [CrossRef]
15. Ma, Z.; Bykova, N.V.; Igamberdiev, A.U. Cell signaling mechanisms and metabolic regulation of germination and dormancy in barley seeds. *Crop J.* **2017**, *5*, 459–477. [CrossRef]

16. Chandrasekaran, U.; Liu, A. Stage-specific metabolization of triacylglycerols during seed germination of Sacha Inchi (*Plukenetia volubilis* L.). *J. Sci. Food Agric.* **2014**, *95*, 1764–1766. [CrossRef]

17. Wu, F.; Yang, N.; Chen, H.; Jin, Z.; Xu, X. Effect of Germination on Flavor Volatiles of Cooked Brown Rice. *Cereal Chem.* **2011**, *88*, 497–503. [CrossRef]

18. Wattanakul, N.; Morakul, S.; Lorjaroenphon, Y.; Na Jom, K. Integrative metabolomics-flavoromics to monitor dynamic changes of 'Nam Dok Mai' mango (*Mangifera indica* Linn) wine during fermentation and storage. *Food Biosci.* **2020**, *35*, 100549. [CrossRef]

19. Kamjijam, B.; Bednarz, H.; Suwannaporn, P.; Jom, K.N.; Niehaus, K. Localization of amino acids in germinated rice grain: Gamma-aminobutyric acid and essential amino acids production approach. *J. Cereal Sci.* **2020**, *93*, 102958. [CrossRef]

20. Gracka, A.; Jeleń, H.H.; Majcher, M.; Siger, A.; Kaczmarek, A. Flavoromics approach in monitoring changes in volatile compounds of virgin rapeseed oil caused by seed roasting. *J. Chromat.* **2016**, *1428*, 292–304. [CrossRef] [PubMed]

21. Atala, E.; Vásquez, L.; Speisky, H.; Lissi, E.; López-Alarcón, C. Ascorbic acid contribution to ORAC values in berry extracts: An evaluation by the ORAC-pyrogallol red methodology. *Food Chem.* **2009**, *113*, 331–335. [CrossRef]

22. Chirinos, R.; Zorrilla, D.; Aguilar-Galvez, A.; Pedreschi, R.; Campos, D. Impact of roasting on fatty acids, tocopherols, phytosterols, and phenolic compounds present in Plukenetia huayllabambana seed. *J. Chem.* **2016**, *3*, 1–10. [CrossRef]

23. Brand-Williams, W.; Cuvelier, M.E.; Berset, C. Use of a free radical method to evaluate antioxidant activity. *LWT-Food Sci. Tech.* **1995**, *28*, 25–30. [CrossRef]

24. Re, R.; Pellegrini, N.; Proteggente, A.; Pannala, A.; Yang, M.; Rice-Evans, C. Antioxidant activity applying an improved ABTS radical cation decolorization assay. *Free Rad. Biol. Med.* **1999**, *26*, 1231–1237. [CrossRef]

25. Gu, E.J.; Kim, D.W.; Jang, G.J.; Song, S.H.; Lee, J.I.; Lee, S.B.; Kim, B.M.; Cho, Y.; Lee, H.J.; Kim, H.J. Mass-based metabolomic analysis of soybean sprouts during germination. *Food Chem.* **2017**, *217*, 311–319. [CrossRef]

26. Shu, X.-L.; Frank, T.; Shu, Q.-Y.; Engel, K.-H. Metabolite Profiling of Germinating Rice Seeds. *J. Agric. Food Chem.* **2008**, *56*, 11612–11620. [CrossRef]

27. Han, C.; Yin, X.; He, D.; Yang, P. Analysis of proteome profile in germinating soybean seed, and its comparison with rice showing the styles of reserves mobilization in different crops. *PLoS ONE* **2013**, *8*, e56947. [CrossRef]

28. Pritchard, S.L.; Charlton, W.L.; Baker, A.; Graham, I.A. Germination and storage reserve mobilization are regulated independently in Arabidopsis. *Plant J.* **2002**, *31*, 639–647. [CrossRef] [PubMed]

29. Boa, F.G.; McDonnell, E.M.; Wilkinson, M.C.; Laidman, D.L. Free sterols and glycolipids in the aleurone tissue of germinating wheat. *Phytochemistry* **1984**, *23*, 519–524. [CrossRef]

30. Yiming, Z.; Hong, W.; Linlin, C.; Xiaoli, Z.; Wen, T.; Xinli, S. Evolution of nutrient ingredients in tartary buckwheat seeds during germination. *Food Chem.* **2015**, *186*, 244–248. [CrossRef] [PubMed]

31. Chen, L.; Wu, J.E.; Li, Z.; Liu, Q.; Zhao, X.; Yang, H. Metabolomic analysis of energy regulated germination and sprouting of organic mung bean (Vigna radiata) using NMR spectroscopy. *Food Chem.* **2019**, *286*, 87–97. [CrossRef]

32. Yang, P.; Li, X.; Wang, X.; Chen, H.; Chen, F.; Shen, S. Proteomic analysis of rice (*Oryza sativa*) seeds during germination. *Proteomics* **2007**, *7*, 3358–3368. [CrossRef]

33. Huang, Y.; Cai, S.; Ye, L.; Hu, H.; Li, C.; Zhang, G. The effects of GA and ABA treatments on metabolite profile of germinating barley. *Food Chem* **2016**, *192*, 928–933. [CrossRef]

34. Wang, K.; Arntfield, S.D. Effect of protein-flavour binding on flavour delivery and protein functional properties: A special emphasis on plant-based proteins. *Flav. Frag. J.* **2017**, *32*, 92–101. [CrossRef]

35. Guichard, E. Interactions between flavor compounds and food ingredients and their influence on flavor perception. *Food Rev. Int.* **2002**, *18*, 49–70. [CrossRef]

36. Paravisini, L.; Guichard, E. Interactions between aroma compounds and food matrix. *Food Percep.* **2016**, 208. [CrossRef]

37. Kumar, V.; Rani, A.; Tindwani, C.; Jain, M. Lipoxygenase isozymes and trypsin inhibitor activities in soybean as influenced by growing location. *Food Chem.* **2003**, *83*, 79–83. [CrossRef]

38. Kaneko, S.; Kumazawa, K.; Nishimura, O. Studies on the key aroma compounds in soy milk made from three different soybean cultivars. *J. Agric. Food Chem.* **2011**, *59*, 12204–12209. [CrossRef]

39. Azarnia, S.; Boye, J.I.; Warkentin, T.; Malcolmson, L. Changes in volatile flavour compounds in field pea cultivars as affected by storage conditions. *Int. J. Food Sci. Technol.* **2011**, *46*, 2408–2419. [CrossRef]

40. Ruan, Y.; Cai, Z.; Deng, Y.; Pan, D.; Zhou, C.; Cao, J.; Chen, X.; Xia, Q. An untargeted metabolomic insight into the high-pressure stress effect on the germination of wholegrain *Oryza sativa* L. *Food Res. Int.* **2021**, *140*, 109984. [CrossRef] [PubMed]

41. Murat, C.; Bard, M.-H.; Dhalleine, C.; Cayot, N. Characterisation of odour active compounds along extraction process from pea flour to pea protein extract. *Food Res. Int.* **2013**, *53*, 31–41. [CrossRef]

42. Kaczmarska, K.T.; Chandra-Hioe, M.V.; Frank, D.; Arcot, J. Aroma characteristics of lupin and soybean after germination and effect of fermentation on lupin aroma. *LWT* **2018**, *87*, 225–233. [CrossRef]

43. Pramai, P.; Abdul Hamid, N.A.; Mediani, A.; Maulidiani, M.; Abas, F.; Jiamyangyuen, S. Metabolite profiling, antioxidant, and α-glucosidase inhibitory activities of germinated rice: Nuclear-magnetic-resonance-based metabolomics study. *J. Food Drug Anal.* **2018**, *26*, 47–57. [CrossRef] [PubMed]

44. Kiritsakis, A.K. Flavor components of olive oil—A review. *J. Am. Oil Chem. Soc.* **1998**, *75*, 673–681. [CrossRef]

45. Gan, R.-Y.; Wang, M.-F.; Lui, W.-Y.; Wu, K.; Corke, H. Dynamic changes in phytochemical composition and antioxidant capacity in green and black mung bean (Vigna radiata) sprouts. *Food Sci. Tech. Res.* **2016**, *51*, 2090–2098. [CrossRef]

46. Shohag, M.J.I.; Wei, Y.; Yang, X. Changes of folate and other potential health-promoting phytochemicals in legume seeds as affected by germination. *J. Agric. Food Chem.* **2012**, *60*, 9137–9143. [CrossRef] [PubMed]
47. Ti, H.; Zhang, R.; Zhang, M.; Li, Q.; Wei, Z.; Zhang, Y.; Tang, X.; Deng, Y.; Liu, L.; Ma, Y. Dynamic changes in the free and bound phenolic compounds and antioxidant activity of brown rice at different germination stages. *Food Chem.* **2014**, *161*, 337–344. [CrossRef] [PubMed]
48. Gan, R.-Y.; Lui, W.-Y.; Wu, K.; Chan, C.-L.; Dai, S.-H.; Sui, Z.-Q.; Corke, H. Bioactive compounds and bioactivities of germinated edible seeds and sprouts: An updated review. *Trends Food Sci. Technol.* **2017**, *59*, 1–14. [CrossRef]
49. Gunenc, A.; Rowland, O.; Xu, H.; Marangoni, A.; Hosseinian, F. Portulaca oleracea seeds as a novel source of alkylresorcinols and its phenolic profiles during germination. *LWT* **2019**, *101*, 246–250. [CrossRef]
50. Chu, C.; Du, Y.; Yu, X.; Shi, J.; Yuan, X.; Liu, X.; Liu, Y.; Zhang, H.; Zhang, Z.; Yan, N. Dynamics of antioxidant activities, metabolites, phenolic acids, flavonoids, and phenolic biosynthetic genes in germinating Chinese wild rice (*Zizania latifolia*). *Food Chem.* **2020**, *318*, 126483. [CrossRef]

Article

Determination of the Personal Nutritional Status of Elderly Populations Based on Basic Foodomics Elements

Natalija Uršulin-Trstenjak [1,2], Ivana Dodlek Šarkanj [1,*], Melita Sajko [2], David Vitez [3] and Ivana Živoder [2]

1   Department of Food Technology, University Center Koprivnica, University North, Trg dr. Žarka Dolinara 1, 48000 Koprivnica, Croatia; natalija.ursulin-trstenjak@unin.hr
2   Department of Nursing, University Center Varaždin, University North, Jurja Križanića 31b, 42000 Varaždin, Croatia; msajko@unin.hr (M.S.); izivoder@unin.hr (I.Ž.)
3   County Hospital Čakovec, I.G.Kovačića 1e, 40000 Čakovec, Croatia; david.vitez.7@gmail.com
*   Correspondence: idsarkanj@unin.hr

Abstract: Nutritional status is a series of related parameters collected using different available methods. In order to determine the nutritional status of elderly populations and ensure nutritional support based on an individual approach, the implementation of the increasingly used foodomics approach is available; this approach plays a key role in personalized diets and in the optimization of diets for a population group, such as an elderly population. The Mini Nutritional Assessment (MNA) method and the Nottingham Screening Tool (NST) form were used on 50 users in a home for the elderly in northwest Croatia. A loss of body mass (BM) was statistically significantly higher for those who had the following: decreased food intake in the last week and users who had complete and partial feeding autonomy. Additionally, the obtained data on drug intake, fluid, individual nutrients, and physical activity are based on an individual approach. The available documentation provides insight into nutritional values and food preparation in an attempt to satisfy a holistic approach in the evaluation of exposure while trying to achieve as many elements of foodomics as possible.

Keywords: elderly population; individual nutritional status assessment; elderly homes; exposure; foodomics

Citation: Uršulin-Trstenjak, N.; Dodlek Šarkanj, I.; Sajko, M.; Vitez, D.; Živoder, I. Determination of the Personal Nutritional Status of Elderly Populations Based on Basic Foodomics Elements. *Foods* **2021**, *10*, 2391. https://doi.org/10.3390/foods10102391

Academic Editors: Yelko Rodríguez Carrasco and Bojan Šarkanj

Received: 19 August 2021
Accepted: 6 October 2021
Published: 9 October 2021

Publisher's Note: MDPI stays neutral with regard to jurisdictional claims in published maps and institutional affiliations.

1. Introduction

People over 65 years of age require a routine nutritional status assessment at least once a year, while those older than 75 years require more frequent assessments [1]. Numerous indicators are used in such assessments, such as anthropometric measurements, laboratory parameters, clinical examination, function tests, and questionnaires, all of which provide a guaranteed approach to each user/respondent. Foodomics plays a crucial role in personalized diets and the optimization of diets for groups of the population, an example of which is the elderly. The increasing frequency of impaired nutritional status has prompted the innovation of simple and effective patterns as a screening to detect both malnutrition and obesity among the general, hospitalized, and institutionalized older population. These screenings contain questions about inadvertent weight loss (further body mass—BM) and body mass index (BMI), dietary habits, and functional status. They represent an insight into the need for either the introduction of nutritional support or the use of an appropriately prescribed diet [2,3].

The four questions in each pattern relate to a loss of BM over the last three months, food intake, BMI, and disease progression. The British Society for Parenteral and Enteral Nutrition and its experts put these issues together and use them in the Nottingham Screening Tool (NST). New and more comprehensive nutritional assessment forms applicable in hospitals, elderly homes, and the community have emerged from the following: (1) Subjective Global Assessment (SGA), which is a fast, widely available, and inexpensive technique with good reproducibility; (2) a validated method, namely the Mini Nutritional

Assessment (MNA); (3) a universal form, i.e., the Malnutrition Universal Screening Tool (MUST) containing a five-step assessment algorithm; and (4) a nutritional risk assessment tool—Nutritional Risk Screening (NRS 2002), which was developed and validated as a questionnaire by the European Society for Clinical Nutrition and Metabolism (ESPEN) [3,4].

Impaired nutritional status adversely affects physical and mental health, resulting in increased complications, higher mortality rates, and increased treatment costs. Malnutrition patients are more susceptible to infection and other health-related complications. In healthy older individuals, malnutrition decreases quality of life and increases the risk of sarcopenia and bone fragility [5], while obese people are exposed to cardiovascular disease, diabetes, and renal dysfunction [6–9]. An essential problem in assessing nutrient intake its interaction with antinutrients [10], heavy metals [11], mycotoxins [12–15], and pesticides [16] that, depending on their size [17–19], can be found in macroscopic clusters or nanoparticles. Thus, a more holistic approach for assessing exposures has recently been advocated, one that estimates intakes and interactions between particular food components [20]. All of these phenomena, together with inherent properties assessed by whole-genome sequencing (WGS) or single-nucleotide polymorphism analysis (SNP), can help in creating personalized nutrition plans that, combined with foodomics, can, in theory, provide the best results for preserving the health of older populations.

Considering that users in elderly homes are a population with long-standing habits of consuming a particular, traditional diet, it is essential to balance and meet the needs of these users through technological preparation and the processing of food, respecting the energy and nutritional value of a given meal [21]. Considering that this is a sensitive population, much attention should be paid to the implementation of a hazard analysis and a critical control points (HACCP) system to ensure the health and safety of the technological preparation of a meal while respecting ethical manufacturing and hygiene practices [22,23]. Additionally, the highest priority should be the respect of users' wishes and preferences, which are emphasized in the ESPEN guideline; this is particularly important in homes for elderly individuals [4].

Foodomics encompasses analytical platforms for researching food composition and, with this, its nutritional values and impact on health are demonstrable. New techniques also provide insight into the detailed picture of food quality and can be used to discover food deceits, as well as to find solutions for other challenges that arise in food production. Foodomics was introduced at an international conference in 2009 in Cesena, Italy, and it was defined as "a discipline that studies areas of food and diet by using and integrating advanced –omics technologies with the purpose of advancing the well-being, health and knowledge of consumers". It requires a combination of food chemistry, biological sciences, and analysis of dates. It encompasses the four main areas of omics: genomics, transcriptomics, proteomics, and metabolomics [24]. Generally, the results of Foodomics's research have a direct impact on consumers, the food industry, and society at large [24]. Regardless of the fact that foodomics is a relatively new research discipline, it is developing fast and expectations related to its growth are high. It is believed that it will become an effective tool for the development of healthy food that is adapted to individual health challenges, and that it can help in preventing illnesses connected to food and food deceits [25,26].

2. Materials and Methods

This research aims to apply a form for assessing nutritional status, eating habits, and health status of users/respondents (after this referred to as users) in elderly homes. We seek to obtain insights not only into the nutritional value of prepared meals, but also regarding the intake and interaction between individual portions of food by assessing exposures, as well as the application of the HACCP system, which ensures the correct preparation of meals in these institutions. The survey included 50 users in elderly homes in the Republic of Croatia, and it was conducted between October and December 2019 with the consent of their institutions.

Through the research, which was conducted between October and December 2019, 50 users of nursing homes in the Republic of Croatia have been surveyed with the approval of the institution. This research was conducted using tools and forms used for evaluating nutritive status (in hospitals, nursing homes, and in the care for elderly people in the community). Four questions were considered in each of the forms, and they were related to a loss of BM in the last three months, BMI, food consumption, and the advancement of illness. According to experts from the British Association for Parenteral and Enteral Nutrition, those questions are connected in the Nottingham Screening Tool-NST. As well as these data, the questionnaire (Questionnaire S1) also included general data about the examinee: sex, age, body weight, body height, diagnosis, and ITM [2]. New and more extensive forms for the evaluation of nutritive status have been developed out of this form and these were also used in this research. Those forms are the following:

Subjective Global Assessment—SGA—is a widely used and cheap technic that connects data from a clinical examination and the anamnesis of an illness, through which a quick assessment of nutritive status can be obtained and, if needed, a quick intervention can be provided. Insight is obtained regarding a loss of BM, a change in food consumption, relevant gastrointestinal symptoms, functional status, and metabolic needs of a patient. Occurrence of infection, length of hospital stay, serum albumin and transferrin, muscle strength, and other objective parameters of the SGA method all influence the obtained data [2].

Mini Nutritional Assessment—MNA—is a quick method used with elderly people in hospitals and nursing homes, as well as with people living alone. The goal of this method is the evaluation of the risk of malnutrition and the insurance of early nutritive support. It encompasses anthropometric measurements (BM, body height (BH), scope of the upper arm), short questions for general evaluation (7 questions concerning a loss of body mass, lifestyle, drug intake, and mobility), questions about nutritional intake (8 questions concerning number of meals, food and water intake, and the possibility of autonomous feeding), and a patient's self-assessment of their nutritive and health status [1]. The obtained data are different depending on the conditions under which the MNA is conducted; therefore, this tool is considered to be at its most accurate when conducted with people who live independently from a broader community because of examinee cooperation [2].

Malnutrition Universal Screening Tool—MUST, although it is primarily intended to be used within the elderly population, also has its uses in whole hospital populations. It is comprised of 5 steps, and uses ITM, loss of body mass, and the impact of the degree of the illness on the health of a patient for determining nutritive status [2].

The tool, that is, the Nutritional Risk Screening questionnaire, or NRS, was developed in 2002, and it is comprised of two parts and is recommended by the ESPEN [4]. The first part is comprised of four questions: (1) Is the patient's BMI lower than 20.5? (2) Has the patient inadvertently lost body mass in the last four months? (3) Did the patient have a small food intake in the last week? (4) Is the patient gravely ill? (for example, intensive care, chronic disease) [2].

In case of an affirmative answer to any of those questions, the examiner continues with the second part of the questionnaire that takes a closer look at the evaluation of nutritive status. Once all data are obtained, a classification is established as to whether the patient is exposed to nutritive risk or whether their nutritive status should be monitored once a week [1].

Through such a data collection method, an individual and comprehensive approach is emphasized in developing diets for elderly users in nursing homes; this approach has been applied in this paper.

For nutritional status assessment, we used the Nottingham Screening Tool (NST) questionnaire containing user information (gender, age, body mass (BM), height, diagnosis), body mass index (BMI), data on accidental loss of BM and food intake, and disease severity data, and a rapidly validated method for assessing nutritional status (Mini Nutritional

Assessment, MNA). These tools were used based on anthropometric measurements (BM and body height (BH), upper arm circumference), general assessments (7 questions related to BM loss, lifestyle, medication, and mobility), dietary intake questions (8 questions regarding meal count, food and water intake, and self-feeding options), and patient self-assessment (how the patient perceives his or her nutritional and health status) [1]. Such a way of collecting data is attuned to an individual and comprehensive approach in the creation of a diet for users in elderly home.

Based on the above, an individual questionnaire of 24 questions was created. The first two questions relate to the gender and age of the respondents and the other two to their anthropometric measurements (BM and BH). The following are a series of grouped questions. The first group includes questions about the loss of BM, the number and cause of weight loss, taking medication and lifestyle, and physical activity. The second group includes questions about the dietary intake of the users, namely the number of meals and fluid intake, and the possibility of self-feeding. The third group of questions is about the duration of and recent changes in diet. In a separate question, users' self-assessment of their health status, gastrointestinal symptoms, and metabolic needs were examined in relation to illness. The research described in this paper shows a correlation between patients' health status and their nutritional status (Figure S1).

The following software packages were used for statistical data processing and graph creation: Microsoft Excel 2016 (Microsoft, Redmond, WA, USA), Statistica 13.3 (TIBCO Software Inc., Palo Alto, CA, USA), Tableau Desktop 2020.4 (Tableau software, Seattle, WA, USA), and Flourish studio. For statistical tests, the normality of data distribution was examined with Shapiro–Wilks' W test; a Leven test confirmed the homogeneity of variance. Given the results of these tests, the Student's t-test or the Mann–Whitney U test was used for a comparison of the two groups, and comparisons of multiple groups were conducted with ANOVA and Kruskal–Wallis ANOVA, respectively. The results were considered statistically significantly different when $p \leq 0.05$.

3. Results

3.1. Distribution of Users According to BMI

The BMI calculation formula provides an index for all 50 survey users categorized as nutritional categories. The data obtained show a statistically normal distribution (Figure 1).

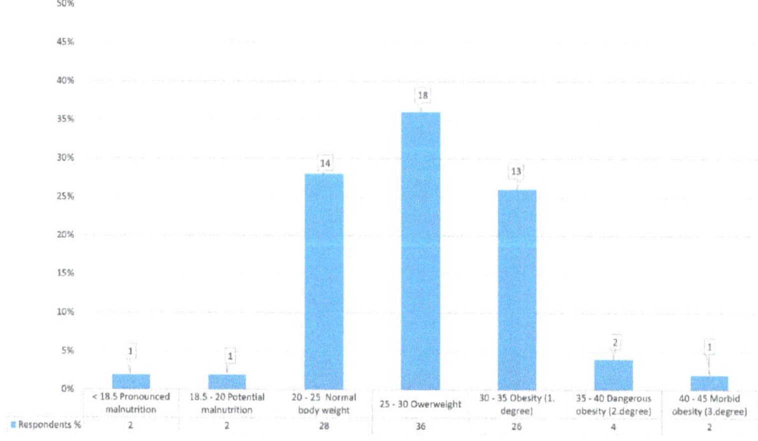

Figure 1. Distribution of users according to body mass index (BMI) shown in absolute and relative terms.

3.2. Loss of BM in Consumer Group Considering Food Intake and Users in Terms of the Possibility of Consuming a Meal

The loss of BM within a one-month period was statistically significantly higher in those with reduced food intake in the previous week (Mann–Whitney U = 43.5; $p < 0.001$) (Figure 2). There was also a statistically significant difference in the distribution of data between users with complete autonomy and users with partial autonomy in feeding ($p = 0.002$).

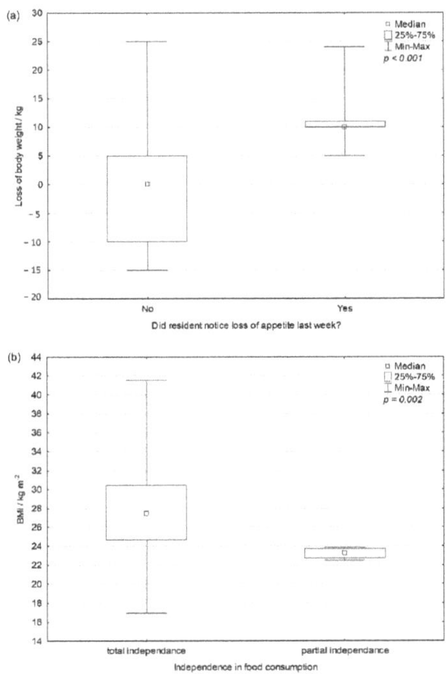

Figure 2. Loss of body mass: (**a**) comparison of user groups concerning food intake; and (**b**) comparison of users about the possibility of consuming meals. The groups were compared by Mann–Whitney U test and shown to be statistically different (p values are on graphs).

3.3. Comparison of the Number of Daily Meals and BM, Common BM, BH, and BMI

Comparing the number of meals that users consume (Figure 3) shows that there is a statistically significant difference between the number of daily meals and BM ($p = 0.041$), the usual BM ($p = 0.021$), and BH ($p = 0.037$), but interestingly not with BMI ($p = 0.334$). BH is well correlated (r = 0.5604; $p < 0.001$) with BM, which is why there is no significant difference in BMI.

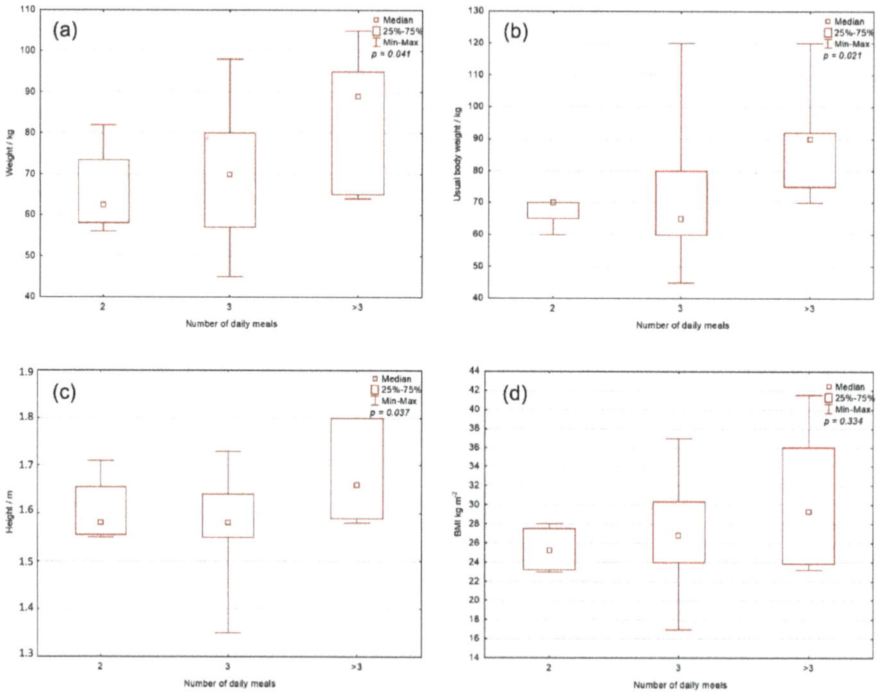

Figure 3. Comparison of the number of daily meals and: (**a**) body mass; (**b**) common body mass; (**c**) height; and (**d**) body mass index. The groups were compared by Mann–Whitney U test (*p* values are on graphs).

3.4. Distribution of Users According to BMI Groups and Gastrointestinal Symptoms

Comparing users divided into BMI groups (Figure 4), 69.5% of participants do not have any gastrointestinal symptom problems in each BMI group, while constipation is the most frequently reported gastrointestinal symptom in the group characterized by dangerous obesity; nausea in the morbid obesity group; and diarrhea in the normal and excessive BM group.

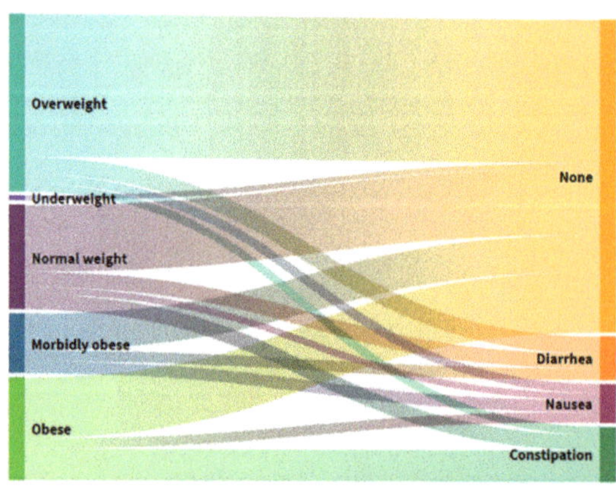

Figure 4. Alluvial diagram of the distribution of users according to the BMI groups and gastrointestinal symptoms.

3.5. Distribution of Users by the Proportion of BMI Groups and Self-Reported Cause of Weight Loss

Figure 5 shows the distribution of the stated mass loss concerning self-assessment.

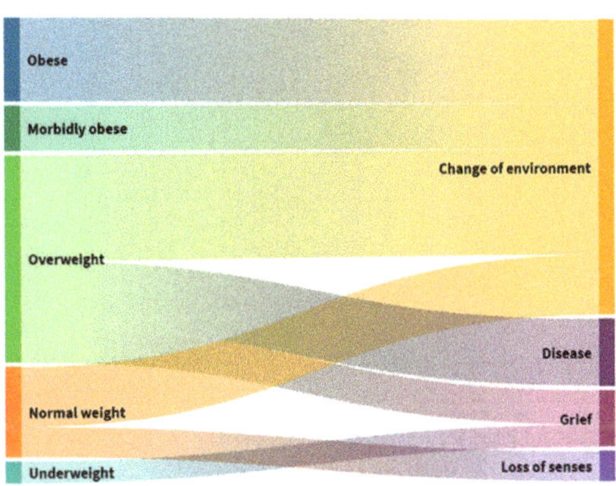

Figure 5. Alluvial diagram of the distribution of users by individual BMI groups and self-reported weight loss cause.

3.6. Distribution of Users According to Individual BMI Groups and Consumption of More Than Three Different Medicines

Users (Figure 6) in the morbidly obese group consumed more than three drugs, the most among all groups. Fewer than three drugs were taken in the group with normal and excessive BM.

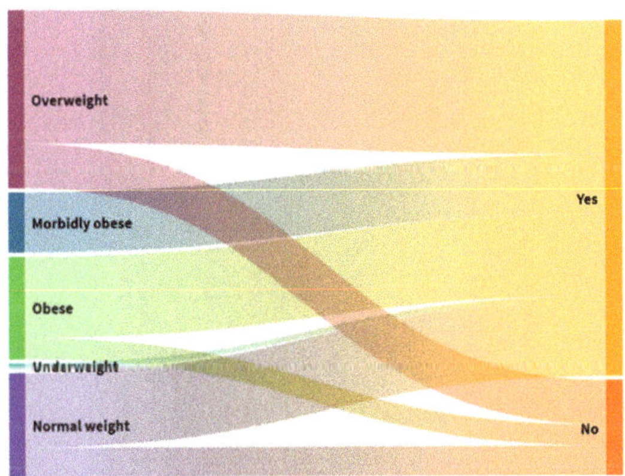

Figure 6. Alluvial diagram of the distribution of users by individual BMI groups and consumption of more than three different drugs.

3.7. Distribution of Users by Individual Groups and Fluid Intake

Figure 7 shows that users with reduced and normal BM take less than 1 L, and those with elevated and excessive BM more than 1 L.

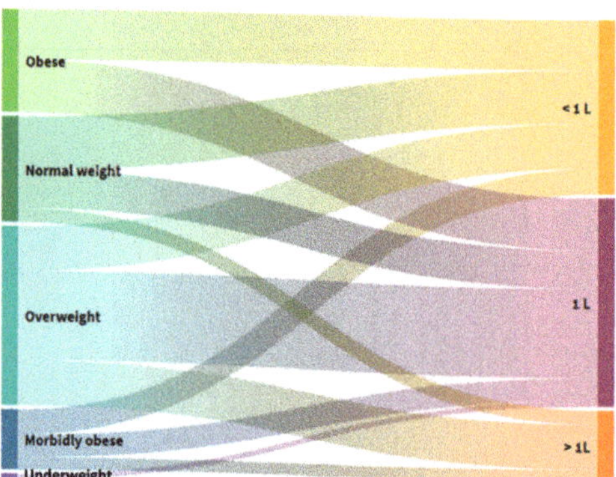

Figure 7. Alluvial diagram of the distribution of users by BMI groups and fluid intake.

3.8. Distribution of Users According to the Frequency of Intake of Macronutrients, Milk and Milk Products, Vegetables, and Fruits Absolutely and Relatively

In the Figure 8 distribution of user's macronutrients, milk, and milk products, vegetables, and fruits are shown. Major proportion are consuming milk and milk products, and carbohydrates every day, while proteins and fats, vegetables and fruits are consumed mainly several time a week.

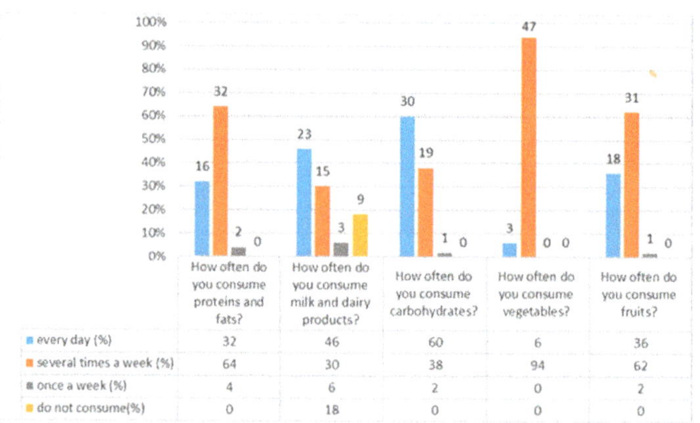

Figure 8. Distribution of users by frequency of intake of macronutrients, milk, and milk products, vegetables, and fruits absolutely and relatively.

3.9. Comparison of the Reduction in Intake of Each Food Group with the Loss of BM Concerning BMI Group

The labels on the Figure 9. show the following: >5 kg—users lost more than 5 kg; 5 kg—users lost 5 kg; <5 kg—users lost less than 5 kg; 0 kg—users did not lose weight. Users also reported changes in their diet and the elimination of or reduction in consumption of a particular group of foods.

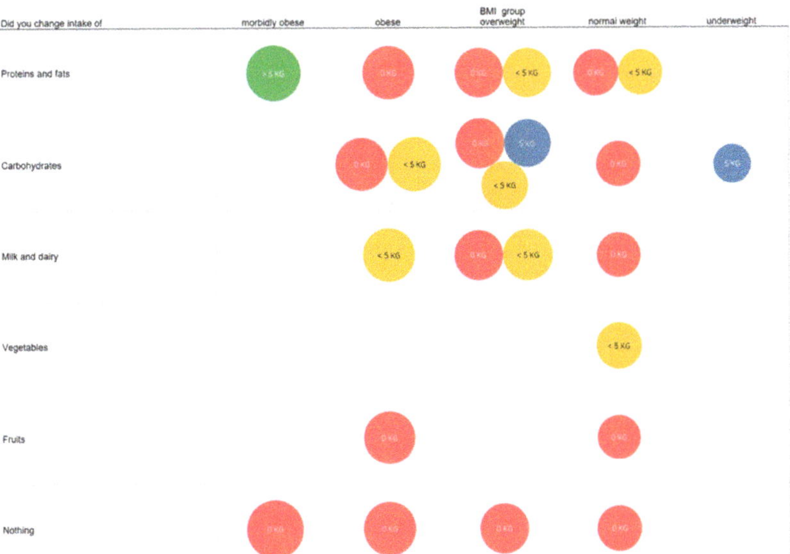

Figure 9. Comparison of dietary intake reduction with weight loss concerning BMI group. The color of bubbles is coded according to weight loss (red—0 kg loss; yellow—up to 5 kg loss; blue—5 kg loss; green—more than 5 kg loss).

3.10. Distribution of Users by BMI Groups by Frequency of Intake of Macronutrient, Milk and Dairy Products, Vegetables, and Fruits

The Figure 10. is divided down by the frequency of food entries from the heading of the graph, using the following abbreviations: ED—every day; MPT—multiple times per week; OPW—once per week; NC—not consuming.

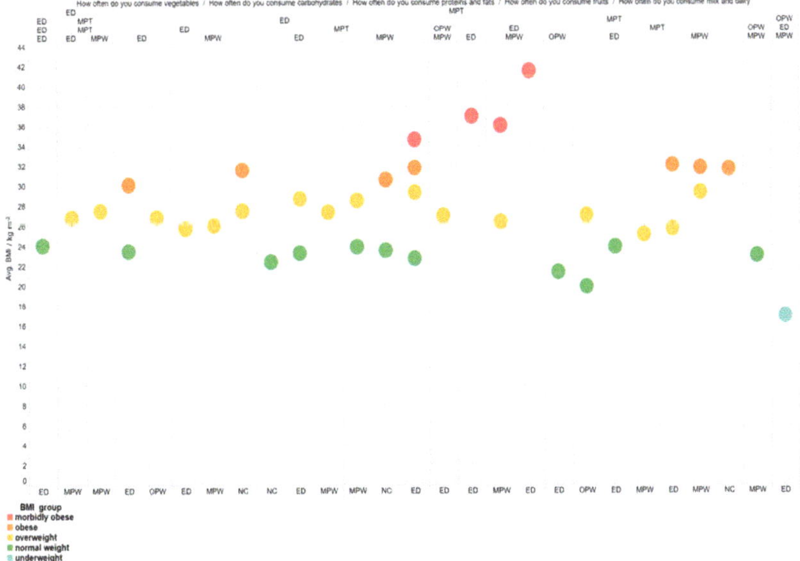

Figure 10. Distribution of users by BMI groups by frequency of intake of macronutrient, milk and dairy products, vegetables, and fruits.

3.11. Distribution of Users according to BMI Groups and Physical Activity (Alone or with Help)

Everyone with morbid obesity (Figure 11), dangerous obesity, and excessive BM engaged in some physical activity, while users with normal BM avoided sports to the fullest.

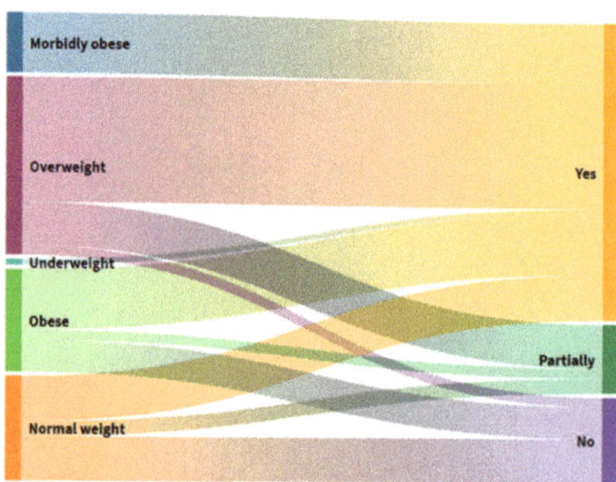

Figure 11. Alluvial diagram of the distribution of users according to the BMI groups and physical activity (alone or with the help).

4. Discussion

This study shows that, out of all the users, three (6%) confirmed that therapy affected BM loss, going on to list common gastrointestinal symptoms such as nausea and diarrhea, while 47 (94%) stated that treatment did not affect their weight loss (Figure S2). More than 250 medicines can affect the absorption and excretion of nutrients. Other side-effects of drugs adversely affecting nutritional status are anorexia (acetylcholinesterase inhibitors, antibiotics, digoxin, hypnotics), early satiety (anticholinergics, sympathomimetics), reduced feeding ability (sedatives, opiates), dysphagia (Nonsteroidal anti-inflammatory drugs (NSAIDs)), constipation (diuretics), and diarrhea (laxatives, antibiotics) [27]. This research was conducted with 50 respondents/users from an elderly home and included 10 (20%) men and 40 (80%) women. According to BMI values and nutritional status, they are classified into nutritional categories with absolute and relative representation. Of the total, only one user was highly malnourished, with a BMI < 18.5 kg/m^2, or 16.94 kg/m^2. One user with a BMI < 20 kg/m^2 and 19.88 kg/m^2 also had potential malnutrition. There were 14 users with normal BMI, (28%) from 20 kg/m^2 to 25 kg/m^2. Eighteen users (36%) were overweight, exceeding the BMI limits of 25 kg/m^2 to 30 kg/m^2. That means that 13 (26%) users are in Grade I obesity and 2 (4%) are in Grade II. One user is in Grade III, which signifies morbid obesity with a BMI > 40 kg/m^2 (Figure 1). In Croatia, the situation is better when compared to a Korean study from 2016 that observed malnutrition in 31% of its examinees, while 49% were at risk of malnutrition and 16% had a normal nutritive status [28]. Additionally, results obtained in Germany from 188 examinees in two elderly homes confirmed that 57.4% were at risk of malnutrition and 15.4%, were malnourished; only 27.1% had a normal nutritive status [29]. The difference in obtained values can be explained by considering that the number of examinees in the mentioned and compared examples was higher. It can also be explained with the use of different tools for the evaluation of the nutritive status within elderly populations and the different degrees of care in different research undertakings [30].

Research has shown that appetite declines with age, as demonstrated by a comprehensive survey by Giezenaar et al. [30]. Although the loss of BM over the observation period provides essential information about the nutritional status of individuals, it is also crucial to consider the percentage of BM loss over the observation period. A loss of 5% indicates a mild and >10% a severe nutritional and health disorder [2].

Wirth et al. show that BM loss is an indicator of protein–energy malnutrition and that it increases the mortality rate of older adults in elderly homes, while BMI < 20 kg/m^2 and weight loss > 5 kg in one year are independent and equally important risk factors for 6-month mortality. An age of 65, BMI < 20 kg/m^2, and weight loss > 5 kg in one year are independent and equally relevant risk factors for the 6-month mortality of elderly individuals [29].

A study by Ryan et al. confirmed that older people who lost at least 5% of their total BM in a month were 4.6 times more likely to die within a year. BM is therefore a useful identifier for mortality in elderly populations. It is essential to note the data obtained from the results of the study, which show that 16 subjects (32%) were obese and their BMI > 30 [30], which approximately corresponds to the data of this research. Both malnutrition and obesity represent serious health risks and medical disorders, and great caution is needed when treating them [2].

Tracking BM loss over a long period is essential information for assessing nutritional status. Failure in the short term is primarily indicative of a disturbed balance in body fluids. In contrast, BM loss over a longer time period indicates changes in metabolism and a decrease in total tissue mass, which puts users at higher nutritional risk. Two (4%) users of this study lost TM within one month, the remaining two (4%) within two months, and nine (18%) over three months. As many as 37 users (74%) suffered no TM loss over three months (Figure S1).

When compared with the research by Ryan et al., whose data show a decrease in BM in 24/153 users (15%), this research shows a loss in only 4% (2 users) in the first month, which is much lower. Therefore, a regular application of the simple anthropometric measurement of BM can be used for detecting users at a higher risk of malnutrition, or even death. The need for further research is apparent in order to observe the role of nutrition in older populations regarding long-term care [31].

Those with reduced dietary intake in the week prior to start of investigation had statistically significantly higher BM Loss (Mann–Whitney U = 43.5; $p < 0.001$) (Figure 2). At the same time, there is a statistically significant difference in the distribution of data between users who have complete autonomy and users who have partial autonomy in feeding ($p = 0.002$). Users with full independence have been found to have higher data dispersal due to eating habits, which can be at both extremes from too much to too little food intake. In those who do not have complete independence, people who help them consume better regulate the quantities and types of foods, so this subgroup is within the boundaries of normal BMI.

Comparing the number of meals that users consume (Figure 3) shows that there is a statistically significant difference between the number of daily meals and BM ($p = 0.041$), the usual BM ($p = 0.021$), and BH ($p = 0.037$), but interestingly not with BMI-a ($p = 0.334$). BH is correlated well (r = 0.5604; $p < 0.001$) with BM, which is why there is no significant difference in BMI. Additionally, users with larger BMs and BHs typically consume more meals per day.

Constipation is common in geriatric populations due to decreased motor function of the colon, reduced fluid intake, and reduced food volume. This study shows that 4 (8%) users often had nausea, 5 (10%) diarrhea, 6 (12%) suffered from constipation, and 35 (70%) had no gastrointestinal symptoms (Figure S2). In terms of the division of users by BMI (Figure 4), most have no problems in each group. In contrast, constipation is most prevalent in the group with dangerous obesity, nausea in the group with morbid obesity, and diarrhea in the group with regular and excessive BM. In elderly homes in other countries, about 60–80% of users have constipation symptoms, which can lead to greater negative consequences and ultimately affect quality of life. Although a high prevalence of constipation in the elderly may be found, there is a lack of empirical evidence to provide interventions based on individual risk factors for constipation [31].

In six people (12%) (Figure S2) who frequently suffered from constipation, a loss of BM was observed in the past three months, citing changes in the environment and loss of taste and odor as a result of self-assessment (Figure 5). Changing the environment as a cause of BM loss is found in all those with elevated BMI and most with normal body weight. Excessive BM users are the second most common cause of BM loss, citing illness and, ultimately, the loss of a loved one. Users with normal BM, with damage caused by a change in the middle, also declared a loss of taste and smell as the main reason for their loss of BM.

The results of this study (Figure S3) show that as many as 39 (78%) users take more than three types, while only 11 (22%) take fewer than three types of medicine. This comparison (Figure 6) shows that everyone in the morbidly obese group belongs to the group that consumes more than three drugs, as do most of the other groups. Those who consume fewer than three medicines were predominantly found in the group with normal and excessive BM. These results confirm that older people suffer more from chronic diseases and take more varied drugs. Results of a study by Morais et al. in 2013, conducted on 664 older people in Europe, have shown that 50.2% of respondents take three or more medicines per day [32], which is, compared to this research, certainly fewer.

Medications can cause a disturbance in food intake by acting directly or indirectly on metabolism, and they can potentially endanger oral health. Older people commonly use anticholinergics, tricyclic antidepressants, sedatives, antihypertensives, muscle relaxants, and benzodiazepines. These drugs reduce the production of saliva, which contributes to a loss of weight and BM and also compromises nutritional status [33].

Older adults often do not consume enough fluids during the day, with the reason being that, with age, the thirst mechanism decreases and they do not feel the need for it even though it is severely deficient in the body [34]. Regarding the users of this study, 23 (46%) take up to 1 L of fluid per day, 19 (38%) < 1 L daily, and only 8 (16%) > 1 L daily (Figure S4), which in no way follows the recommended guidelines [1]. An insight into the further result obtained (Figure 7) shows that users with reduced and typical BM take < 1 L water, and those with elevated and excessive BM > 1 L water. Studies in elderly homes reveal that dehydration is common, and 20–30% of older people suffer from dehydration caused by loss of water; elderly individuals displaying dehydration-related confusion and disorientation recover either entirely or significantly once their fluid intake becomes adequate [35,36]. One prospective study found dehydration in 38.3% of elderly home users, showing that 30.5% of users are at risk of developing dehydration. Total water intake averages 1l to 1.5l, and 96% of users do not meet the estimated daily intake of fluid [37]. When compared to our research, it can be seen that the indicators of inadequate fluid intake are the same, with only 16% intaking > 1 L. In another study, fluid intake was inversely related to an increase in age, cognitive impairment, food consumption challenges, and increased dining staff, and as many as 85% of elderly home users consumed less than 1.5 L of fluid per day [38]. Other studies also confirm the low daily fluid intake of elderly home users. Reed et al. found that 51.3% of nursing home users in the United States, with an average age of 85, consumed less than 237 mL of fluid per day, with as many as 37% of those experiencing severe cognitive impairment [39]. Inadequate total fluid intake and, consequently, dehydration, prevail in the elderly population in nursing homes in all BMI categories. Nutritional interventions should be directed towards giving more attention to the type of liquid beverages, as well as diets in terms of milk, yogurt, and soup.

The reduced protein–energy status of a person causes severe health problems and affects nutritional status. This study sought to obtain as much information as possible on the intake of proteins, fats, milk, dairy products, carbohydrates, and vegetables and fruits through daily diets [8] by looking at menus (Table S1), and has made following the recommendations regarding guidelines [1]. The daily advice for protein consumption is 0.8 g/kg up to 1.5 g/kg, which can influence the prevention of muscle mass loss, i.e., sarcopenia [1]. It is evident from the survey results that 32 (64%) users consumed protein and fat several times a week, 16 (32%) every day, and only 2 (4%) once a week (Figure 8). Similar research was conducted on elderly home users who were monitored for protein status and protein intake regulated by daily diet and enteral preparations. Users had higher BMI and less pronounced symptoms of musculoskeletal pain [40].

Fat consumption in elderly homes largely depends on the type of food and how it is prepared. Research conducted in elderly homes has shown that very little is invested in the selection of quality foods with fewer saturated fatty acids, which poses a risk to the health of elderly populations. Menus in elderly homes contain very few quality sources of unsaturated fatty acids [41]; the recommended intake of total fat is 20–35% [1]. Milk and dairy products were consumed daily by 23 (46%) users, 15 (30%) several times a week, while only 3 (6%) once a week (Graph 8). Insufficient intake of milk and milk products below the recommended level causes the skeletal system to weaken. The daily calcium intake in the elderly population is 1200 mg/day, which is the same as the recommendation for vitamin D, which reduces the risk of complications of osteopenia and osteoporosis [42].

Adequate energy and protein intake could be important in the process of maintaining the health of elderly populations. The data concerning the real food intake of users in nursing homes generally are not yet sufficiently examined.

Despite nationally available UK testing guidelines for menu planning and nutrition for people in elderly homes, as many as fourteen studies show that the food offered does not meet nutritional needs. Another the survey concludes that inadequate intake of milk and dairy products is also associated with a lack of protein intake and an increased risk of malnutrition [43], which has also been shown by Van Zwien et al. (2019) [41] in their study, which states 18% of users achieved a protein intake of 68 g/day, which is less than in this current paper (46%). Croatian national guidelines point out that 50% of the elderly do not have the right amount of vitamin D due to insufficient exposure to the sun, thin skin, and reduced intake of milk, dairy products, and meat [1].

In total, 30 users (60%) in this study consume carbohydrates daily, 19 (38%) multiple times a week, and only one respondent (2%) once a week (Figure 8). The guidelines do not show traditional allowable UH intake values. Still, there is a consensus that an intake preference provided will provide 55–60% of total daily energy intake in the form of compound UHs [1]. Frequent carbohydrate intake per day significantly increases the caloric value of foods. It often causes adverse effects on the psychological function of older individuals, leading to a rapid increase in BMI and a decrease in cognitive function, especially memory [44]. A literature source shows that, in general, carbohydrate-rich foods represent the largest share in almost every single meal. The highest risk for the health of the elderly is the intake of high-calorie carbohydrates, such as refined sugars through cakes, and it has been shown that the consumption of these foods increases mortality and creates a higher incidence of heart disease and stroke in the elderly [7,41].

Consumption of vegetables by the users of this study shows that 3 (6%) consume vegetables every day, and 47 (94%) several times a week. Regarding fruit, 18 (36%) respondents consume it daily, 31 (62%) consume it several times a week, while 1 (2%) eat it once a week (Figure 8). A similar study was conducted by experts on the consumption of fruits and vegetables in a sample of 445 people over 65 years old, and the results have shown that 37% of people living in urban areas and 51% of people in rural areas do not consume the recommended five units of fruits and vegetables daily. In terms of reasoning, participants cited their inferior social status relative to the general population, decreased appetite, and taste sensation [45]. Research findings in other countries showed that few people consumed fruits or vegetables every day [46].

Exceptional care is given to nutrition in both the preparation and nutritional value of food. Food in the home is prepared in three primary meals and two entrees of the local cuisine, respecting variety and being tastefully prepared. Meals tailored to the health needs of users are also prepared: for people with diabetes, low fat and salt meals, vegetarian meals, the ability to adjust according to user preferences, mixed food served in the dining room, or tablet system in rooms (Table S1).

Users of this study self-reported changes in their diet and the elimination of or reduction in consumption of a particular group of foods. After comparing the dietary changes with the loss of BM (Figure 9), it was found that people who reduced protein and fat intake had the highest result in reducing BM. On average, the reduction in carbohydrate intake was the best. People who reduced their intake of fruit or did not reduce their consumption did not show a decrease in BM, unlike in research Morais et al., 2013 [33] showing that 53% of examinees developed nutritive risk by decreasing their consumption of fruit and vegetables (for easier chewing). It is important to draw attention to life circumstances, changes in appetite or health, general perceptions of health, intake of food and vegetables, and choice of food that is easy to chew, as all of these elements influence BMI.

Figure 10 shows how, given the consumption of a particular macronutrient source, users were divided into BMI groups. It follows that respondents who consumed vegetables daily were generally in groups with lower BMIs than others. The group with the highest incidence of morbid obesity consumed proteins and fats, carbohydrates, and vegetables several times a week.

A loss of muscle mass, resulting in a loss of strength, is also one of the essential components of the fragility of older individuals of both sexes. Moderate physical activity is recommended among older people, as it is important to strengthen muscle mass and increase strength and endurance [1]. In this study, 40 (80%) users were physically active, while only 10 (20%) were not (Figure S5). A comparison of the users of this study in different BMI groups (Figure 11) has shown that everyone with morbid obesity engages in some physical activities, thus showing that they care about their health. Additionally, most subjects with dangerous obesity and overweight play sports on their own or with help. Unfortunately, the majority of people who are a healthy weight avoid sports, which could lead them to have problems related to accumulating excess weight in later life; thus, it is recommended that the general population be motivated as much as possible to play sports. Research at the European level shows four clusters with an association between energy–protein intake and physical activity among the elderly; this is the best strategy for achieving adequate body weight [47].

The results of similar research in elderly homes link fragility with physical activity, suggesting that home users have a higher propensity for sedentary lifestyles [48]. According to the study, one hour of immobility per day confirmed the increased risk of sarcopenia by 33% [49]. A sedentary lifestyle in the elderly is associated with decreased functional ability and an increased risk of disability in daily activities [50,51]. Improper food intake can affect the functional status of older people in nursing homes. At the same time, results of a presentation of nutritional status among users in Spanish homes show an inadequate composition of the nutritional value of meals [52]. Older users who do not consume adequate energy input may not have the energy or power to access food and water on their own, request additional food or water, or feed on their own. In a sample of 98 nursing home users, it was confirmed that malnutrition predicted deterioration in functional status [53,54].

The kitchen of the elderly home in question is equipped with the latest food preparation machines and equipment, all following the HACCP system (Figure S6); this is also a legal obligation in the Republic of Croatia. It ensures the implementation of ethical manufacturing and good hygiene practices that guarantee quality technological processes and healthy and safe products. In this way, nutrient intake and its interactions with antinutrients [10], heavy metals [11], (myco) toxins [12–15], and pesticides [16] are monitored; thus, a holistic approach to the assessment of exosomes (which has advocated lately) is conducted [20,55].

This research aims to apply a form for assessing nutritional status, eating habits, and health status of users/respondents in elderly homes.

Through the process of aging, more physical and psychological changes (psychosocial and socioeconomical) are developed, which often influence nutritional needs and nutritive status. In this day and age, chronic diseases are often accompanied by therapy consumption (medicine), which can lead to an imbalance in diet and can often result in a poor nourishment status [56]. Therefore, one of the key elements regarding the care of older individuals is the regular evaluation of their nutritive status. A characteristic of the diet of this population is that their need for energy and macronutrients becomes lower, while their need for micronutrients stays the same or becomes higher than during adulthood. Therefore, the organizing of meals is exceptionally important [57].

The data of a large study encompassing 60 hospitals in the evaluation of the nutritive status of patients through the use of different forms/tools, using SGA, showed that 63.3% of examinees were malnourished, while the application of NRS showed that 90% of hospitalized elderly people were malnourished [58].

Therefore, another goal of this paper is to obtain insight into the foodomics approach, which plays a key role in personalized diets and the optimization of diets for older individuals.

This paper sought to obtain insights not only regarding the nutritional value of prepared meals, but also in terms of the intake and interaction between individual portions of food by assessing exposures, as well as through the application of the HACCP system to ensure that meals were prepared correctly in these institutions.

5. Conclusions

In terms of normal BM, 14 (28%) users have a BMI ranging from 20 kg/m^2 to 25 kg/m^2. Eighteen users (36%) are overweight, which exceeds the BMI limits of 25 kg/m^2 to 30 kg/m^2. This means that 13 (26%) users are in Grade I obesity, and two (4%) are in Grade II; meanwhile, one user is in Grade III, which signifies morbid obesity with a $\text{BMI} > 40 \text{ kg/m}^2$

Users with reduced dietary intake in the week prior to investigation had statistically significantly higher BM loss (Mann–Whitney $U = 43.5; p < 0.001$).

At the same time, there is a statistically significant difference in the distribution of data between users who have complete autonomy and users who have partial autonomy in feeding ($p = 0.002$); users with full independence have been found to have higher data dispersal due to their eating habits.

Comparing the number of meals that users consume shows that there is a statistically significant difference between the number of daily meals and BM ($p = 0.041$), the usual BM ($p = 0.021$), and BH ($p = 0.037$), but interestingly not with BMI- a ($p = 0.334$). BH is correlated well ($r = 0.5604; p < 0.001$) with BM, which is why there is no such significant difference in BMI. Additionally, users of larger BMs and BHs typically consume more meals per day.

This research highlights the importance of checking and continuously monitoring the nutritional status of those in elderly homes to prevent nutritional risk. It is also essential to ensure the preparation of a nutritionally valuable healthy meal by applying and implementing an HACCP system in addition to technologically traditional forms of food processing that seek to satisfy a holistic approach to assessing exposures. Furthermore, the use of an individual and comprehensive approach in elderly homes has to be emphasized in order to ensure adequate nutritional intake of food and liquids, to improve the health status of users, and to advance the quality of life of elderly people.

All of these elements are connected in foodomics as a tool for the development of healthy food that is adapted to individual health challenges and the prevention of illnesses connected to food in order to improve the increasingly longer lifetime of humans.

Diet problems, which often result in malnutrition, are generally spread among elderly people in nursing homes. The prevalence rates differ depending on the parameters and limit values used for evaluating nutrition, as well as on the examined population. Future studies should carefully characterize their participants and use standardized tools for evaluating nutrition in order to achieve a better comparison of their results.

Supplementary Materials: The following are available online at https://www.mdpi.com/article/10.3390/foods10102391/s1, Supplementary file 1—Content: Questionnaire S1: An individualized questionnaire prepared according to the guidelines of standardized methods for assessing the nutritional status of the elderly for research purposes, Figure S1: Distribution of users according to weight loss over a period of 1 month, 2 months and 3 months in absolute and relative terms, Figure S2: Review of the user's response to a question about the presence of gastrointestinal symptoms in absolute and relative terms, Figure S3: Review of user's response to a question about taking medication and the effect of therapy on weight loss in absolute and relative terms, Figure S4: Distribution of users according to fluid intake daily in absolute and relative terms, Figure S5: Review user's response to a question: Are you physically active? in absolute and relative terms, Figure S6: certification of HACCP system certification. Table S1: The menu of nursing home used for research.

Author Contributions: Conceptualization, N.U-T.; methodology, I.D.Š.; software, I.D.Š.; validation, N.U.-T. and I.D.Š.; formal analysis, I.D.Š.; investigation, D.V. and M.S.; resources, N.U.-T. and M.S.; data curation, N.U.-T., I.D.Š., M.S. and D.V.; writing—original draft preparation, N.U.-T., M.S. and I.Ž.; writing—review and editing, N.U.-T., I.D.Š., M.S. and I.Ž.; visualization, N.U.-T.; supervision, N.U.-T.; project administration, N.U.-T.; funding acquisition, N.U.-T. All authors have read and agreed to the published version of the manuscript.

Funding: The research was funded by "Aid for Scientific Research and Art Work of the University of the North-Biotechnical Area of Science".

Informed Consent Statement: Informed consent was obtained from all subjects involved in the study.

Conflicts of Interest: The authors declare no conflict of interest.

References

1. Krznarić, Ž.; Vranešić Bender, D.; Ljubas Kelečić, D.; Reiner, Ž.; Tomek Roksandić, S.; Kekez, D.; Pavić, T. Croatian Guidelines for Nutrition in The Elderly, Part I (Croatian). *Liječnički Vjesnik.* **2011**, *133*, 231–240.
2. Štimac, D.; Krznarić, Ž.; Vranešić Bender, D.; Obrovac Glišić, M. *Dietotherapy and Clinical Nutrition*; Medicinska naklada: Zagreb, Croatia, 2014; pp. 26–33.
3. BAPEN. Available online: https://www.bapen.org.uk/screening-and-must/must/must-toolkit (accessed on 19 February 2020).
4. Volkert, D.; Beck, A.M.; Cederholm, T.; Cruz-Jentoft, A.; Goisser, S.; Hooper, L.; Kiesswetter, E.; Maggio, M.; Raynaud-Simon, A.; Sieber, C.C.; et al. ESPEN guideline on clinical nutrition and hydration in geriatrics. *Clin. Nutr.* **2019**, *38*, 10–47. [CrossRef] [PubMed]
5. Martone, A.M.; Onder, G.; Vetrano, D.L.; Ortolani, E.; Tosato, M.; Marzetti, E.; Landi, F. Anorexia of Aging: A Modifiable Risk Factor for Frailty. *Nutrients* **2013**, *5*, 4126–4133. [CrossRef] [PubMed]
6. Uršulin-Trstenjak, N.; Kosalec, N.; Levanić, D. Complementary feeding of infants in northwest Croatia. *J. Hyg. Eng. Des.* **2018**, *24*, 44–54.
7. Uršulin-Trstenjak, N.; Levanić, D.; Hasaković-Felja, M. Pretilost kao faktor rizika za nastajanje kardiovaskularnih-koronarnih bolesti. *Tech. J.* **2015**, *9*, 230–234.
8. Uršulin-Trstenjak, N.; Vitez, B.; Levanić, D.; Sajko, M.; Neuberg, M. Protein-energy status in patients receiving dialysis. *J. Hyg. Eng. Des.* **2017**, *19*, 26–30.
9. Slobođanac, M.; Uršulin-Trstenjak, N. *Prehrana Kroničnih Bubrežnih Bolesnika*; Polytechnic in Varaždin: Varaždin, Croatia, 2010; pp. 13–19.
10. Thompson, L.U. Potential health benefits and problems associated with antinutrients in foods. *Food Res. Int.* **1993**, *26*, 131–149. [CrossRef]
11. Hejna, M.; Gottardo, D.; Baldi, A.; Dell'Orto, V.; Cheli, F.; Zaninelli, M.; Rossi, L. Review: Nutritional ecology of heavy metals. *Animal* **2018**, *12*, 2156–2170. [CrossRef]
12. Ezekiel, C.N.; Sulyok, M.; Ogara, I.M.; Abia, W.A.; Warth, B.; Šarkanj, B.; Turner, P.C.; Krska, R. Mycotoxin sin uncooked and plate-ready household food form rural northern Nigeria. *Food Chem. Toxicol.* **2019**, *128*, 171–179. [CrossRef]
13. Kovač, M.; Šubarić, D.; Bulaić, M.; Kovač, T.; Šarkanj, B. Yesterday masked, today modified; what do mycotoxins bring next? *Arch. Hig. Rada. Toksikol.* **2018**, *69*, 196–214. [CrossRef]
14. Šarkanj, B.; Ezekiel, C.N.; Turner, P.C.; Abia, W.A.; Rychlik, M.; Krska, R.; Sulyok, M.; Warth, B. Ultra-sensitive, stable isotope assisted quantification of multiple urinary mycotoxin exposure biomarkers. *Anal. Chim. Acta* **2018**, *1019*, 84–92. [CrossRef]
15. Wang, Q.; Šarkanj, B.; Jurasović, J.; Chisti, Y.; Sulyok, M.; Gong, J.; Sirisansaneeyakul, S.; Komes, D. Evaluation of microbial toxins, trace elements and sensory properties of a high-thea brownins instant Pu-erh tea produced using Aspergillus tubingensis via submerged fermentation. *Int. J. Food Sci. Technol.* **2019**, *54*, 1541–1549. [CrossRef]
16. Reeves, W.R.; McGuire, M.K.; Stokes, M.; Vicini, J.L. Assessing the safety of pesticides in food: How current regulations protected human health. *Adv. Nutr.* **2019**, *10*, 80–88. [CrossRef] [PubMed]
17. Kovač, T.; Šarkanj, B.; Klapec, T.; Borišev, I.; Kovač, M.; Nevistić, A.; Strelec, I. Antiaflatoxigenic effect of fullerene C60 nanoparticles at environmentally plausible concentrations. *AMB Expr.* **2018**, *8*, 1–8. [CrossRef]
18. Kovač, T.; Borišev, I.; Crevar, B.; Čačić Kenjerić, F.; Kovač, M.; Strelec, I.; Ezekiel, C.N.; Sulyok, M.; Krska, R.; Šarkanj, B. Fullerol C60(OH)24 nanoparticles modulate aflatoxin B1 biosynthesis in Aspergillus flavus. *Sci. Rep.* **2018**, *8*, 12855. [CrossRef] [PubMed]

19. Kovač, T.; Borišev, I.; Kovač, M.; Lončarić, A.; Čačić Kenjerić, F.; Đorđev, A.; Strelec, I.; Ezekiel, C.N.; Sulyok, M.; Krska, R.; et al. Impact of fullerol C60(OH)24 nanoparticles on the production of emerging toxins by Aspergillus flavus. *Sci. Rep.* **2020**, *10*, 725. [CrossRef] [PubMed]
20. Warth, B.; Spangler, S.; Fang, M.; Johnson, C.N.; Forsberg, E.M.; Grenados, A.; Martin, R.L.; Domingo-Almenara, X.; Huan, T.; Rinehart, D.; et al. Exposome-scale investigation guided by global metabolomics, pathway analysis, and cognitive computing. *Anal. Chem.* **2017**, *89*, 11505–11513. [CrossRef]
21. Uršulin-Trstenjak, N.; Levanić, D.; Šalamon, D.; Puntarić, D.; Sajko, M. Implementation of informatization in organizing hospital meals in the Republic of Croatia. *J. Hyg. Eng. Des.* **2016**, *17*, 18–23.
22. Uršulin-Trstenjak, N.; Vahčić, N.; Šušnić, S. Food safety system through the HACCP system in catering. In Proceedings of the 2008 Joint Central European Congres, 4th Central European Congress of Food, 6th Croatian Congress of Food Technologistic, Biotechnologists and Nutritionists, Cavtat, Croatia, 15–17 May 2008; Čurić, D., Ed.; Croatian Chamber of Economy: Zagreb, Croatia, 2008; pp. 165–168.
23. Uršulin-Trstenjak, N.; Levanić, D.; Šušnić, S.; Gazibara, D.; Šušnić, V. Implementation of the system and application software for the monitoring of the food safety system in hospitality. *Tech. J.* **2014**, *8*, 166–170.
24. Capozzi, F.; Bordoni, A. Foodomics: A new comprehensive approach to food and nutrition. *Genes Nutr.* **2013**, *8*, 1–4. [CrossRef]
25. Clish, C. Metabolomics: An emerging but powerful tool for precision medicine. *Cold Spring Harb. Mol. Case Studies* **2015**, *1*, a000588. [CrossRef]
26. Cifuentes, A. *Foodomics: Food Science & Omics Tools in the 21st Century*; Foodomics Laboratory: Madrid, Spain, 2015.
27. Ship, J.A.; Pillemer, S.R.; Baum, B.J. Xerostomia and the Geriatric Patient. *J. Am. Geriatr. Soc.* **2002**, *50*, 535–543. [CrossRef]
28. Wakabayashi, H.; Matsushima, M. Dysphagia assessed by the 10-item eating assessment tool is associated with nutritional status and activities of daily living in elderly individuals requiring long-term care. *J. Nutr. Health Aging.* **2016**, *1*, 22–27. [CrossRef] [PubMed]
29. Giezenaar, C.; Chapman, I.; Luscombe-Marsh, N.; Feinle-Bisset, C.; Horowitz, M.; Soenen, S. Ageing Is Associated with Decreases in Appetite and Energy Intake—A Meta-Analysis in Healthy Adults. *Nutrients* **2016**, *8*, 28. [CrossRef] [PubMed]
30. Wirth, R.; Streicher, M.; Smoliner, C.; Kolb, C.; Hiesmayr, M.; Thiem, U.; Volkert, D. The impact of weight loss and low BMI on mortality of nursing home residents—Results from the nutrition Day in nursing homes. *Clin. Nutr.* **2016**, *35*, 900–906. [CrossRef] [PubMed]
31. Ryan, C.; Bryant, E.; Eleazer, P.; Rhodes, A.; Guest, K. Unintentional weight loss in long-term care: A predictor of mortality in the elderly. *South Med. J.* **1995**, *88*, 721–724. [CrossRef]
32. Huang, T.T.; Yang, S.D.; Tsai, Y.H.; Chin, Y.F.; Wang, B.H.; Tsay, P.K. Effectiveness of individualized intervention on older residents with constipation in a nursing home: A randomized controlled trial. *J. Clin. Nurs.* **2015**, *24*, 3449–3458. [CrossRef] [PubMed]
33. Morais, C.; Oliveira, B.; Afonso, C.; Lumbers, M.; Raats, M.; Almedia, M.D.V. Nutritional risk of European elderly. *Eur. J. Clin. Nutr.* **2013**, *67*, 1215–1219. [CrossRef] [PubMed]
34. Zadak, Z.; Hyspler, R.; Ticha, A.; Vlcek, J. Polypharmacy and malnutrition. *Curr. Opin. Clin. Nutr. Metab. Care* **2013**, *16*, 51–55. [CrossRef] [PubMed]
35. Touhy, A.T.; Jett, K.F. *Gerontological Nursing and Healthy Aging*, 4th ed.; Elsevier Mosby: St. Louis, MO, USA, 2014; pp. 141–143.
36. Hooper, L.; Bunn, D.; Jimoh, F.O.; Fairweather-Tait, S.J. Water-loss dehydration and aging. *Mech. Ageing Dev.* **2014**, *136–137*, 50–58. [CrossRef] [PubMed]
37. Begum, M.N.; Johnson, C.S. A review of the literature on dehydration in the institutionalized elderly. *E-SPEN Eur. E J. Clin. Nutr. Metab.* **2010**, *5*, e47–e53. [CrossRef]
38. Marra, M.V.; Simmons, S.F.; Shotwell, M.S.; Hudson, A.; Hollingsworth, E.K.; Long, E.; Silver, H.J. Elevated Serum Osmolality and Total Water Deficit Indicate Impaired Hydration Status in Residents of Long-Term Care Facilities Regardless of Low or High Body Mass Index. *J. Acad. Nutr. Diet* **2016**, *116*, 828–836. [CrossRef] [PubMed]
39. Namasivayam-MacDonald, A.M.; Slaughter, S.E.; Morrison, J.; Steele, C.M.; Carrier, N.; Lengyel, C.; Keller, H.H. Inadequate fluid intake in long term care residents: Prevalence and determinants. *Geriatr. Nurs.* **2018**, *39*, 330–335. [CrossRef] [PubMed]
40. Reed, P.S.; Zimmerman, S.; Sloane, P.D.; Williams, C.S.; Boustani, M. Characteristics associated with low food and fluid intake in long-term care residents with dementia. *Gerontologist* **2005**, *45* (Suppl. 1), 74–80. [CrossRef]
41. Van Zwienen-Pot, J.I.; Visser, M.; Kruizenga, H.M. Predictors for achieving adequate protein and energy intake in nursing home rehabilitation patients. *Aging Clin. Exp. Res.* **2017**, *30*, 799–809. [CrossRef] [PubMed]
42. Saava, M.; Kisper-Hint, I.R. Nutritional assessment of elderly people in the nursing house and at home in Tallinn. *J. Nutr. Health Aging* **2002**, *6*, 93–95.
43. National Institute of Health. Available online: https://www.bones.nih.gov/health-info/bone/bone-health/nutrition/calcium-and-vitamin-d-important-every-age (accessed on 19 February 2020).
44. Iuliano, S.; Poon, S.; Wang, X.; Bui, M.; Seeman, E. Dairy food supplementation may reduce malnutrition risk in institutionalised elderly. *Br. J. Nutr.* **2017**, *117*, 142–147. [CrossRef] [PubMed]
45. Young, K.W.H.; Greenwood, C.E.; van Reekum, R.; Binns, M.A. A Randomized, Crossover Trial of High-Carbohydrate Foods in Nursing Home Residents with Alzheimer's Disease: Associations Among Intervention Response, Body Mass Index, and Behavioral and Cognitive Function. *J. Gerontol. A Biol. Sci. Med. Sci.* **2005**, *60*, 1039–1045. [CrossRef]

46. Johnson, A.E.; Donkin, A.J.M.; Morgan, K.; Neale, R.J.; Page, R.M.; Silburn, R.L. Fruit and vegetable consumption in later life. *Age Ageing* **1998**, *27*, 723–728. [CrossRef]
47. Soini, H.; Routasalo, P.; Lagström, H. Characteristics of the Mini-Nutritional Assessment in elderly home-care patients. *Eur. J. Clin. Nutr.* **2004**, *58*, 64–70. [CrossRef]
48. Yung Hung, Y.; Wijnhoven, H.A.H.; Visser, M.; Verbeke, W. Appetite and Protein Intake Strata of Older Adults in the European Union: Socio-Demographic and Health Characteristics, Diet-Related, and Physical Activity Behaviours. *Nutrients* **2019**, *11*, 777. [CrossRef]
49. Reid, N.; Eakin, E.; Henwood, T.; Keogh, J.W.; Senior, H.E.; Gardiner, P.A.; Winkler, E.; Healy, G.N. Objectively measured activity patterns among adults in residential aged care. *Int. J. Environ. Res. Public Health* **2013**, *10*, 6783–6798. [CrossRef]
50. Gianoudis, J.; Bailey, C.A.; Daly, R.M. Associations between sedentary behavior and body composition, muscle function and sarcopenia in community-dwelling older adults. *Osteoporos. Int.* **2015**, *26*, 571–579. [CrossRef]
51. Santos, D.A.; Silva, A.M.; Baptista, F.; Santos, R.; Vale, S.; Mota, J.; Sardinha, L.B. Sedentary behavior and physical activity are independently related to functional fitness in older adults. *Exp. Gerontol.* **2012**, *47*, 908–912. [CrossRef]
52. Dunlop, D.D.; Song, J.; Arnston, E.K.; Semanik, P.A.; Lee, J.; Chang, R.W.; Hootman, J.M. Sedentary time in US older adults associated with disability in activities of daily living independent of physical activity. *J. Phys. Act. Health* **2015**, *12*, 93–101. [CrossRef] [PubMed]
53. Rodríguez-Rejón, A.I.; Ruiz-López, M.D.; Artacho, R. Dietary Intake and Associated Factors in Long-Term Care Homes in Southeast Spain. *Nutrients* **2019**, *11*, 266. [CrossRef] [PubMed]
54. Knezić, K.; Pahor, Đ.; Pavić, E.; Poljak, V.; Uremović, S.; Vahčić, N.; Vazdar, R.; Vodopija Sušanj, D. *Vodič Dobre Higijenske Prakse i Primjene HACCP Načela za Institucionalne Kuhinje*; Ministarstvo zdravstva: Zagreb, Croatia, 2010.
55. Šarkanj, B.; Dodlek Šarkanj, I.; Shamtsyan, M. Mycotoxins in food—how to prevent and what to do when things go bad. *E3S Web of Conf.* **2020**, *215*, 01004. [CrossRef]
56. Zuliani, G.; Romagnoni, F.; Volpato, S.; Soattin, L.; Leoci, V.; Bollini, M.C.; Buttarello, M.; Lotto, D.; Fellin, R. Nutritional Parameters, Body Composition, and Progression of Disability in Older Disabled Residents Living in Nursing Homes. *J. Gerontol. Biol. Sci. Med. Sci.* **2001**, *56*, 212–216. [CrossRef] [PubMed]
57. Diekmann, R.; Winning, K.; Uter, W.; Kaiser, M.J.; Sieber, C.C.; Volkert, D.; Bauer, J.M. Screening for malnutrition among nursing home residents—a comparative analysis of the mini nutritional assessment, the nutritional risk screening, and the malnutrition universal screening tool. *J. Nutr. Health Aging* **2013**, *17*, 326–331. [CrossRef] [PubMed]
58. Pablo, A.M.; Izaga, M.A.; Alday, L.A. Assessment of nutritional status on hospital admission: Nutritional scores. *Eur. J. Clin. Nutr.* **2003**, *57*, 824–831. [CrossRef]

 *foods*

Article

Screening of Regulated and Emerging Mycotoxins in Bulk Milk Samples by High-Resolution Mass Spectrometry

Gabriele Rocchetti [1,2,]*, Francesca Ghilardelli [1], Francesco Masoero [1] and Antonio Gallo [1]

[1] Department of Animal Science, Food and Nutrition, Faculty of Agricultural, Food and Environmental Sciences, Università Cattolica del Sacro Cuore, 29122 Piacenza, Italy; francesca.ghilardelli@unicatt.it (F.G.); francesco.masoero@unicatt.it (F.M.); antonio.gallo@unicatt.it (A.G.)

[2] Department for Sustainable Food Process, Faculty of Agricultural, Food and Environmental Sciences, Università Cattolica del Sacro Cuore, 29122 Piacenza, Italy

* Correspondence: gabriele.rocchetti@unicatt.it

Abstract: In this work, a retrospective screening based on ultra-high-performance liquid chromatography (UHPLC) coupled with high-resolution mass spectrometry (HRMS) based on Orbitrap-Q-Exactive Focus™ was used to check the occurrence of regulated and emerging mycotoxins in bulk milk samples. Milk samples were collected from dairy farms in which corn silage was the main ingredient of the feeding system. The 45 bulk milk samples were previously analyzed for a detailed untargeted metabolomic profiling and classified into five clusters according to the corn silage contamination profile, namely: (1) low levels of *Aspergillus*- and *Penicillium*-mycotoxins; (2) low levels of fumonisins and other *Fusarium*-mycotoxins; (3) high levels of *Aspergillus*-mycotoxins; (4) high levels of non-regulated *Fusarium*-mycotoxins; (5) high levels of fumonisins and their metabolites. Multivariate statistics based on both unsupervised and supervised analyses were used to evaluate the significant fold-change variations of the main groups of mycotoxins detected when comparing milk samples from clusters 3, 4, and 5 (high contamination levels of the corn silages) with cluster 1 and 2 (low contamination levels of the corn silages). Overall, 14 compounds showed a significant prediction ability, with antibiotic Y (VIP score = 2.579), bikaverin (VIP score = 1.975) and fumonisin B2 (VIP score = 1.846) being the best markers. The k-means clustering combined with supervised statistics showed two discriminant groups of milk samples, thus revealing a hierarchically higher impact of the whole feeding system (rather than the only corn silages) together with other factors of variability on the final mycotoxin contamination profile. Among the discriminant metabolites we found some *Fusarium* mycotoxins, together with the tetrapeptide tentoxin (an *Alternaria* toxin), the α-zearalenol (a catabolite of zearalenone), mycophenolic acid and apicidin. These preliminary findings provide new insights into the potential role of UHPLC-HRMS to evaluate the contamination profile and the safety of raw milk to produce hard cheese.

Keywords: milk metabolomics; retrospective screening; UHPLC-Orbitrap; multivariate statistics; mycotoxins

Citation: Rocchetti, G.; Ghilardelli, F.; Masoero, F.; Gallo, A. Screening of Regulated and Emerging Mycotoxins in Bulk Milk Samples by High-Resolution Mass Spectrometry. *Foods* 2021, 10, 2025. https://doi.org/10.3390/foods10092025

Academic Editors: Yelko Rodríguez-Carrasco and Bojan Šarkanj

Received: 5 August 2021
Accepted: 26 August 2021
Published: 28 August 2021

1. Introduction

Milk is an important constituent of the human diet in the Western world [1]. In recent decades, world milk production has increased by over 60%, from 522 million tonnes in 1987 to 843 million tonnes in 2018 [1]. As recently reviewed [2,3], the spectrum of milk metabolites can be deeply influenced by several external factors, such as the season, origin, health status, processing, storage, formulation, and feeding systems.

In this regard, one of the major concerns when considering the feeding system is related to the potential contamination of silages (such as corn silage) by mycotoxins [4,5]. Mycotoxin contamination, especially in milk, has evoked global concerns regarding feed and food safety due to the toxic effects of mycotoxins in both animals and humans [6,7], including carcinogenic, mutagenic, teratogenic, immunotoxic, and estrogenic potential.

Mycotoxins are secondary metabolites produced by several fungi and mainly belonging to the *Aspergillus*, *Penicillium*, *Fusarium*, *Alternaria*, and *Claviceps* strains [8]. These genera can produce a wide range of different mycotoxins for which specific regulations have been established in many countries to protect consumers and livestock from their harmful effects [5]. The great ingestion by dairy cows of regulated mycotoxins has been related to the quality of the feeding systems [9]; however, this aspect remains insufficiently investigated. The available data trying to connect the quality of forages and silages (such as corn silage) with other kinds of the so-called emerging mycotoxins (characterized by being neither routinely determined nor legislatively regulated) are very scarce [5,10]. When dairy cows receive a mycotoxin-contaminated feed, these fungal secondary metabolites can be metabolized and potentially transferred to animal-derived food, including milk and dairy products [8]. This aspect might be of great concern, considering the toxic effects previously mentioned.

To date, few analytical methods have been proposed for the simultaneous identification of mycotoxins using high-resolution approaches in milk and dairy products [11]. Regarding the most advanced analytical methods available for this purpose, ultra-high-performance liquid chromatography (UHPLC) has overcome the limitations of conventional HPLC (e.g., lower separation capacity and speed of analysis), improving sensitivity and resolution using packing materials with smaller particle size [12]. Although tandem mass spectrometry (MS/MS) provides adequate quantification and high efficiency for multi-residue analyses, this strategy is sometimes limited for analysis at trace levels in complex matrices. High-resolution mass spectrometry (HRMS) using Orbitrap technology has made it possible to achieve high resolution and good specificity because of the mass accuracy provided by the HRMS detector, combined with traditional information [12,13]. This technique also enables the retrospective analysis of samples, in contrast to MS/MS, by using appropriate and/or ad hoc databases [12,14,15].

Starting from these background conditions, in this work, a high-resolution UHPLC-Orbitrap mass spectrometry approach was used to evaluate the mycotoxins profile of bulk milk samples collected from dairy farms using corn silages as the main ingredients of the total mixed ration (TMR) (i.e., 30.51 ± 5.84% on a dry matter, DM, basis) [16]. The corn silages were classified in five main clusters according to the mycotoxin contamination, as previously reported in Gallo et al. [5]. On the same milk samples, we have previously demonstrated the potential of untargeted metabolomics to evaluate the impact of contaminated corn silages on the most important chemical classes, thus providing evidence that sphingolipids, together with purine and pyrimidine-derived metabolites, are the most affected groups of metabolites [16]. Therefore, the aim of this work was to provide new insights into the ability of a retrospective screening based on HRMS to assess the mycotoxin profile of bulk milk samples, thus potentially evaluating food safety issues as related to both regulated and emerging toxins.

2. Materials and Methods

2.1. Collection of Milk Samples

Samples of bulk tank milk (500 mL) (n = 45) were taken in the period January–June 2018 from dairy farms located in the Po Valley (Italy). These latter farmed Holstein Friesian housed in free-stall barns without pasture access [16]. The herds were characterized by having on average 38.3% ± 1.9 of primiparous on lactating dairy cows and 2.4 ± 0.2 lactations before culling. The dairy cows were milked twice a day (i.e., morning and afternoon milking sessions) and fed a diet based on the large use of corn silage. In particular, the corn silage represented the main ingredient of total mixed ration (TMR) (i.e., 30.51 ± 5.84% on a dry matter basis). The visited dairy herds provided additional information when considering other ingredients characterizing the TMR, namely other small-grain silages (i.e., 10.65 ± 6.80% on dry matter basis) and hay (i.e., 10.26 ± 5.98% on dry matter basis). The lactating dairy cows were fed with the same corn silages, previously analyzed by Gallo et al. [5], from at least four weeks, thus avoiding collecting milk in the period in which corn silage bunkers were changed. Additional information regarding herd composition, milk yield of lactating groups and milk quality, dry matter intake (DMI) as well as TMR formulation characteristics are reported in our previous published works [5,16]. The corn silages were grouped in five clusters according to the mycotoxin contamination profiles, as reported in our previous work [5], namely: cluster 1 (corn silages contaminated by low levels of both *Aspergillus*- and *Penicillium*-produced mycotoxins); cluster 2 (corn silages contaminated by low levels of fumonisins, and other *Fusarium*-produced mycotoxins); cluster 3 (corn silages contaminated by high levels of *Aspergillus*-mycotoxins); cluster 4 (corn silages contaminated by high levels of *Fusarium*-produced mycotoxins); cluster 5 (corn silages contaminated by high levels of fumonisins and their metabolites; number of samples: 3). The collected 45 bulk milk samples were then classified according to the same corn silages grouping, thus obtaining the following five groups: 18 samples (cluster 1), 17 samples (cluster 2), 2 samples (cluster 3), 5 samples (cluster 4), and 3 samples (cluster 5). The sample legend for each cluster can be found in Table S1.

2.2. Extraction Step for UHPLC-HRMS Analysis

Milk samples were extracted according to the method previously reported for untargeted screening [16–18]. Following a skimming process by centrifugation ($4500 \times g$ for 10 min at 4 °C), the 45 milk samples ($n = 3$) were thoroughly vortex mixed. Afterwards, an aliquot of 2 mL of each sample was added to 14 mL of acetonitrile (LC-MS grade, Sigma-Aldrich, Madison, CA, USA) acidified with 3% formic acid, mixed by vortexing for 2 min and processed with ultrasounds for 5 min. The samples were centrifuged at $12,000 \times g$ for 15 min at 4 °C to remove large biomolecules (such as proteins). The supernatants were then filtered through 0.22 μm cellulose syringe filters in amber vials until the further untargeted metabolomic screening.

2.3. Screening of Mycotoxins by UHPLC-HRMS Analysis

The UHPLC-HRMS analysis was based on an untargeted metabolomic approach, by using a Q Exactive™ Focus Hybrid Quadrupole-Orbitrap Mass Spectrometer (Thermo Scientific, Waltham, MA, USA) coupled to a Vanquish ultra-high-pressure liquid chromatography (UHPLC) pump and equipped with a HESI-II probe (Thermo Scientific, USA), as previously reported by [5]. The chromatographic separation was achieved under a water-acetonitrile (both LC-MS grade, from Sigma-Aldrich, Milan, Italy) gradient elution (6–94% acetonitrile in 35 min) using 0.1% formic acid as phase modifier, on an Agilent Zorbax Eclipse Plus C18 column (50 × 2.1 mm, 1.8 μm). For the full scan MS analysis, the acquisition was performed using the positive ionization with a mass resolution of 70,000 at m/z 200. The automatic gain control target (AGC target) and the maximum injection time (IT) were $1e^6$ and 100 ms, respectively. Additionally, randomized injections of pooled quality control (QC) samples were performed in a data-dependent (Top N = 3) MS/MS mode with full scan mass resolution reduced to 17,500 at m/z 200, with an AGC target value of $1e^5$, maximum IT of 100 ms, and isolation window of 1.0 m/z, respectively. For the stage of data-dependent MS/MS, the Top N ions were selected for further fragmentation under stepped normalized collisional energy (i.e., 10, 20, 40 eV). The injection volume was 6 μL and the m/z range for the full scan analyses was 100–1200. Heated electrospray ionization (HESI) parameters were as follows: sheath gas flow 40 arb (arbitrary units), auxiliary gas flow 20 arb, spray voltage 3.5 kV, and capillary temperature 320 °C. Prior to data collection, the mass spectrometer was calibrated using a Pierce™ positive ion calibration solution (Thermo Fisher Scientific, San Jose, CA, USA). To avoid possible bias, the sequence of injections for milk samples was randomized. Additionally, blank samples (i.e., extraction solvent only) were randomly injected through the sequence.

2.4. Data Processing

The collected data (.RAW files) were converted into abf format using the Reifycs Abf Converter and then further processed using the software MS-DIAL (version 4.60) [19].

Automatic peak finding, LOWESS normalization, and annotation via spectral matching against the database Mass Bank of North America were initially carried out. The mass range 100–1200 *m/z* was searched for features with a minimum peak height of 10,000 cps. The MS and MS/MS tolerance for peak centroiding were set to 0.01 and 0.05 Da, respectively. Retention time information was excluded from the calculation of the total score. Accurate mass tolerance for identification was 0.01 Da for MS and 0.05 Da for MS/MS. The identification step was based on mass accuracy, isotopic pattern, and spectral matching. In MS-DIAL, these criteria were used to calculate a total identification score. The total identification score cut-off was >50%, considering the most common HESI+ adducts. Gap filling using peak finder algorithm was performed to fill in missing peaks, considering 5 ppm tolerance for *m/z* values. The software MS-Finder [20] was also used for in-silico fragmentation of the non-annotated mass features, using different available data sources, such as FoodDB, BMDB, PubChem, T3DB (Toxin) and KNApSAcK libraries, thus reaching a level 2 of confidence in annotation [21]. A custom database containing the mycotoxins previously identified in the corn silage samples [5] was also used for a tentative annotation according to the accurate mass and isotopic profile of each compound and exploiting the MS-DIAL software.

Finally, calibration curves of authentic standards (purity $\geq$ 97%) of α-zearalenol (CAS number: 36455-72-8), mycophenolic acid (CAS number: 24280-93-1), apicidin (CAS number: 183506-66-3), and tentoxin (CAS number: 28540-82-1) (from Sigma-Aldrich) were injected considering the concentration range: 0.1–100 ng/mL. Data were finally elaborated in the same software to provide semi-quantitative values. In our experimental conditions and according to literature [13,22], the definition of the limit of detection and limit of quantification is not applicable due to the application of high-resolution mass spectrometric method. However, to ensure quantification, a certain degree of confidence is required. Therefore, the limit of quantification for the semi-quantitative analysis was the lowest calibration level used (0.1 ng/mL).

2.5. Multivariate Statistical Analysis

The HRMS data were elaborated using the software MetaboAnalyst [23,24]. Briefly, after data normalization, both unsupervised and supervised multivariate statistics were carried out. The unsupervised approach was based on hierarchical cluster analysis (HCA) and k-means clustering approach, while the orthogonal projections to latent structures discriminant analysis (OPLS-DA) was used as supervised tool. Additionally, the OPLS-DA model validation parameters (goodness-of-fit R^2Y together with goodness-of-prediction Q^2Y) were inspected, considering a Q^2Y prediction ability of >0.5 as the acceptability threshold. Thereafter, the OPLS-DA model produced was inspected for outliers and permutation testing (N > 100) was performed to exclude model over-fitting. The importance of each mycotoxin detected for discrimination purposes was then calculated according to the variable selection method VIP (i.e., variables importance in projection), considering as the minimum significant threshold those values higher than 1, also inspecting the S-plot related to the OPLS-DA model built. As the next step, volcano plots were produced for the comparison between contaminated (cluster 3, 4, and 5) vs. control groups (cluster 1 and 2) by coupling fold-change analysis (cut-off value > 1.2) and ANOVA ($p < 0.05$; post-hoc test: Tukey HSD; multiple testing correction: Bonferroni Family-Wise Error Rate).

3. Results and Discussion

3.1. Profiling of Mycotoxins in the Different Milk Samples

The starting mycotoxin contamination profile of the different corn silage clusters is summarized in Figure 1, considering each major group of mycotoxins detected, as previously discussed in Gallo et al. [5].

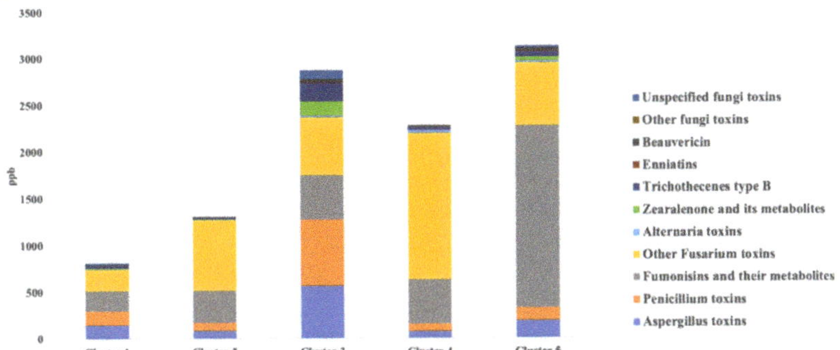

Figure 1. Cumulative values (expressed as µg/kg dry matter) of the major mycotoxins and their metabolites detected in the corn silages belonging to the different clusters.

Overall, silages belonging to cluster 1 and cluster 2 were characterized by the lowest contamination levels (cumulatively lower than 1500 µg/kg), while cluster 3, 4, and 5 were highly contaminated by several regulated and emerging mycotoxins (Figure 1). From a qualitative point of view, the corn silages revealed a similar mycotoxin contamination profile, which was mainly associated with *Aspergillus* toxins, *Penicillium* toxins, fumonisins and their metabolites, together with other *Fusarium* toxins. Additionally, our previous work [16] demonstrated that there was a clear impact of contaminated corn silages on the milk metabolomic profiles, and this was particularly true for the metabolism of purines, pyrimidines, sphingolipids, and oxidative stress-related compounds (such as oxidized glutathione). However, it is important to highlight that in this work no information is available on the contamination of the other TMR ingredients potentially affecting the final contamination profile of milk samples. Therefore, after having evaluated separately the fermentative quality characteristics of corn silages [5] and then the untargeted metabolomic profile of bulk milk from dairy cows consuming those contaminated corn silages [16], we exploited a retrospective screening based on high-resolution mass spectrometry to comprehensively investigate the mycotoxin contamination profile of the same bulk milk samples.

Overall, the retrospective screening following HRMS based data acquisition allowed us to identify 46 mycotoxins and/or metabolites, which are reported in Table S1 considering their adduct type, reference *m/z*, formula, total identification score (as provided by MS-Dial software), MS1 isotopic spectrum, MS/MS spectrum (where available), and relative abundance values for each sample replicate ($n = 3$). In our experimental conditions, the group composed of other *Fusarium* mycotoxins was found to be the most represented in the final dataset, being composed of 13 compounds (such as fusaric acid and apicidin), followed by 7 *Penicillium* mycotoxins, 5 toxins produced by other fungal strains (including ilicicolin A, ilicicolin B, citreorosein, macrosporin, and iso-rhodoptilometrin) and *Alternaria* mycotoxins (such as alternariol and tentoxin). Additionally, among the 46 mycotoxins detected, 10 compounds were structurally confirmed by means of MS/MS annotations, namely 4Z-infectopyron, kojic acid, fumonisin B2, nivalenol, siccanol, culmorin, 15-hydroxyculmorin, butenolide, beauvericin, and pestalotin (Table S1). Interestingly, the HRMS approach revealed a wide distribution of mycotoxins and some of their metabolites in the bulk milk samples under investigation. In our previous work [5], 69 mycotoxins were identified and quantified in the different corn silages. Therefore, from a qualitative point of view, we found that milk samples were characterized by the 63.7% of mycotoxins found in the corn silages, although this latter was the not exclusive ingredient of the TMR of the visited dairy farms. Future ad hoc studies (such as those based on the evaluation of carry-over phenomena and individual diets for the dairy cows) are extremely necessary to better investigate the toxicological risks for both animals and humans.

3.2. Multivariate Analysis on the Different Milk Samples and Discriminant Metabolites

In this work, a foodomics-based approach was used to retrospectively screen the mycotoxins in the different bulk milk samples to produce hard cheese, to find potential marker compounds of the feeding regimen (typically based on corn silage as the main ingredient). Therefore, starting from the mycotoxin profile reported in Table S1 and previously discussed, we used a multivariate statistical approach based on a hierarchical clustering to naively group the five milk clusters according to the mycotoxins detected. In this regard, the resulting heat map (Figure 2) was built considering the average log fold-change (FC) variations of each mycotoxin detected across the five different contamination clusters.

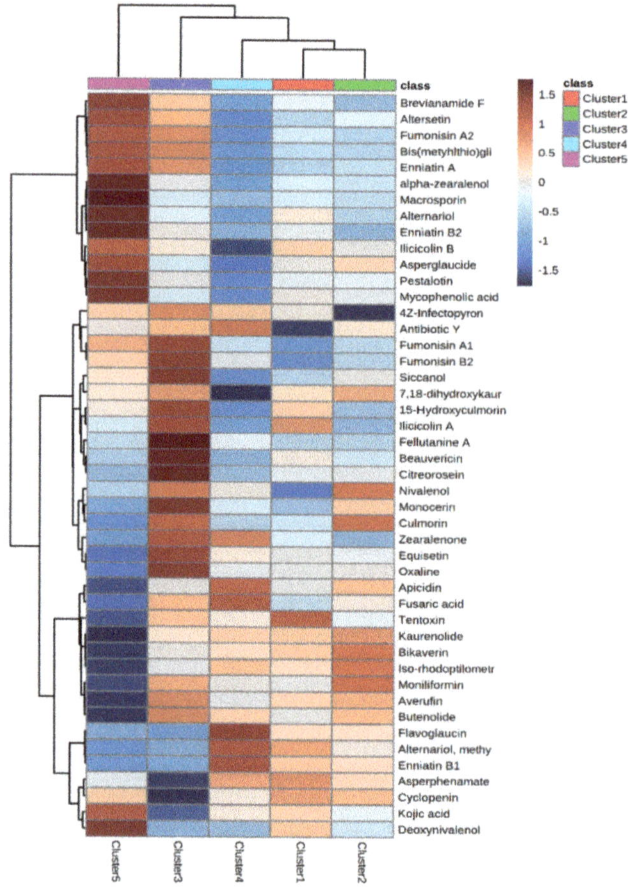

Figure 2. Heat map showing the average clustering (distance measure: Euclidean; clustering algorithm: Ward) of the milk samples according to the mycotoxins detected. The log2 fold-change (FC) values of each abundance were used for the cluster analysis with the Mass Profiler Professional software. Red and blue colors in each column indicate relative up- or down-accumulation of the mycotoxins, respectively.

As can be observed from the figure, the average mycotoxin distribution based on the corn silage clusters allowed us to group milk samples into two main groups: the first group (on the right side of the heat map) consisted of milk samples belonging to the clusters 1 and 2 (from silages with a low contamination level), cluster 4 (from silages with high levels of non-regulated *Fusarium* mycotoxins), and cluster 3 (high levels of *Aspergillus*-mycotoxins in the silages). Additionally, cluster 5 (i.e., high levels of fumonisins and their metabolites) showed the most differential and exclusive profile. The heat map reported in Figure 2 allowed us to observe an indirect correlation between the contamination profile of corn silages and milk samples, as also reported in our previous work evaluating the global changes of the major chemical classes because of the contaminated feeding systems [16]. Interestingly, cluster 5 was outlined as the most discriminant also in our previous work, thus highlighting the potential of high levels of fumonisins and their metabolites in the corn silages to potentially drive chemical differences in cow milk. Additionally, it was evident from the heat map that some mycotoxins were characterized by both strong up (red color) and down (blue color) accumulation trends in the different milk clusters.

The supervised OPLS-DA approach combined with the VIP selection method was then used to extrapolate those mycotoxins characterized by the highest discrimination potential. The OPLS-DA predictive score plot built considering the distribution of the different mycotoxins detected is reported as Figure 3. As can be observed, the 45 milk samples showed a high variability into the score plot space, already observed in the not-averaged unsupervised HCA (not reported), with the orthogonal component explaining the 16.4% of the prediction ability. Additionally, the OPLS-DA model built was characterized by a goodness of prediction lower than 0.5 (i.e., representing the cut-off of acceptability), thus demonstrating a bad correlation between the mycotoxin contamination profile of corn silages and the bulk milk samples, thus resulting in a prediction model not robust enough to be used for discrimination or traceability purposes. This result is likely due to the other ingredients characterizing the TMR. Therefore, further studies are mandatory to better evaluate the final mycotoxin profile detected. As a next step, we extrapolated those mycotoxins better accounting for the differences between the five different milk clusters, using the VIP selection method. Overall, 15 toxins were outlined as the best in terms of discrimination potential (i.e., VIP score > 1), with antibiotic Y (a *Fusarium* mycotoxin) characterized by the best discriminant ability (VIP score = 2.579). Among the VIP markers, the 26.6% consisted of other *Fusarium* mycotoxins, followed by a quite similar numerical distribution (i.e., two toxins per class) for the remaining VIP markers. Among the VIP markers we found also zearalenone (VIP score = 1.358) and its metabolite α-zearalenol (VIP score = 1.098).

Considering the scarce prediction ability resulting from the supervised OPLS-DA modelling, we decided to combine the information provided by volcano plot (i.e., built by coupling a fold-change analysis with a one-way ANOVA) with the VIP discriminant markers, thus assessing the cumulative LogFC variations of the main classes of mycotoxins. An overview of the different compounds (organized in classes) can be found in Table 1. As a general consideration, the mycotoxins annotated showed differential trends according to the average LogFC values. In this regard, when considering the comparisons with clusters 1 and 2 (considered as a control because of the low contamination levels of corn silages), fumonisins showed an average up-accumulation in milk samples belonging to clusters 3 and 5, whilst *Alternaria* mycotoxins showed no averaged differences, although altersetin was particularly abundant in milk samples characterizing cluster 5. Regarding the group composed of *Aspergillus* mycotoxins, bis(methyl-thio)-gliotoxin was highly and significantly abundant in milk samples belonging to cluster 3 and 5, with average LogFC values of 5.83 and 7.24, respectively. Additionally, regarding zearalenone metabolites, we found a significant down-accumulation of zearalenone in cluster 5 with a corresponding strong and significant ($p < 0.05$) increase in α-zearalenol. Overall, cluster 5 was confirmed again to be the most characteristic in terms of mycotoxin profile; accordingly, it also showed a strong down-accumulation for other *Fusarium* mycotoxins for the comparison with both cluster 1 and cluster 2 (being −1.49 and −1.87, respectively). Regarding the other classes of mycotoxins (such as Enniatins-derived or *Penicillium* toxins), no huge variations were detected for the different milk clusters under investigation (Table 1).

Table 1. Mycotoxins detected by HRMS in the different milk samples. Each compound is provided with its VIP score (from OPLS-DA), log2 fold-change (FC) value and p-value (FWER) resulting from the volcano plot analysis.

Class	Compound	p-Value (FWER)	VIP Score (OPLS-DA)	LogFC Cluster 3 vs. Cluster 1	LogFC Cluster 3 vs. Cluster 2	LogFC Cluster 4 vs. Cluster 1	LogFC Cluster 4 vs. Cluster 2	LogFC Cluster 5 vs. Cluster 1	LogFC Cluster 5 vs. Cluster 2
Alternaria mycotoxins	Altersetin	$p > 0.05$	<1	2.151	1.562	−1.466	−2.056	3.343	2.753
	Alternariol	$p > 0.05$	<1	−0.064	0.113	−0.341	−0.163	−0.097	0.080
	Alternariol, methyl-ether	0.0011	<1	−0.810	−0.433	0.139	0.516	−1.273	−0.897
	Tentoxin	0.0495	1.119	−0.158	0.272	−0.492	−0.061	−1.258	−0.827
	4Z-Infectopyron	0.0354	<1	0.298	1.072	0.002	0.774	−0.210	0.562
Aspergillus mycotoxins	Brevianamide F	0.0065	1.287	0.307	0.671	−0.584	−0.220	0.501	0.865
	Kojic acid	$p > 0.05$	<1	−0.542	−0.513	−0.250	−0.222	−0.267	−0.239
	Averufin	$p > 0.05$	<1	0.531	0.129	0.080	−0.321	−0.640	−1.042
	Bis(methylthio)gliotoxin	0.0066	1.102	5.414	6.242	−2.603	−1.775	6.830	7.657
	Asperphenamate	$p > 0.05$	<1	−5.604	−4.700	−1.351	−0.447	−2.126	−1.222
Fumonisins mycotoxins	Fumonisin A1	0.0494	1.667	2.069	1.671	0.412	0.014	1.084	0.687
	Fumonisin A2	0.0024	<1	3.285	4.041	−4.203	−3.447	4.114	4.870
	Fumonisin B2	0.0294	1.846	2.444	1.918	0.858	0.332	1.096	0.570
Zearalenone and metabolites	Zearalenone	0.0176	1.358	0.420	0.632	0.175	0.386	−0.543	−0.332
	α-Zearalenol	0.0378	1.098	0.416	0.666	−1.264	−1.014	2.822	3.072
Trichothecenes	Deoxynivalenol	$p > 0.05$	<1	−0.396	−0.080	−0.530	−0.214	0.00072	0.316
	Nivalenol	$p > 0.05$	1.114	0.198	0.053	−0.029	−0.174	−0.265	−0.410
Other *Fusarium* mycotoxins	Siccanol	$p > 0.05$	<1	0.335	0.303	−0.261	−0.293	−0.214	−0.246
	Monocerin	$p > 0.05$	<1	0.460	0.287	−0.072	−0.246	−0.376	−0.550
	Moniliformin	5.2×10^{-6}	1.012	0.442	−0.187	−0.119	−0.748	−1.359	−1.988
	Equisetin	$p > 0.05$	<1	1.166	2.234	−0.060	1.006	−1.493	−0.426
	Culmorin	$p > 0.05$	<1	0.243	0.071	−0.185	−0.357	−0.467	−0.639
	15-Hydroxyculmorin	$p > 0.05$	<1	0.137	0.352	−0.377	−0.161	−0.376	−0.161
	Butenolide	$p > 0.05$	<1	0.116	0.097	−0.099	−0.118	−0.560	−0.578
	Bikaverin	3.4×10^{-4}	1.975	−1.630	−3.375	−0.734	−2.478	−9.655	−11.400
	Apicidin	$p > 0.05$	<1	0.104	−1.462	3.163	1.596	−4.700	−6.268
	Antibiotic Y	0.0065	2.579	3.395	1.069	0.771	−1.554	2.315	−0.010
	Kaurenolide	0.0024	1.581	−0.178	−0.334	−0.198	−0.354	−1.845	−2.001
	7,1-Dihydroxykaurenolide	$p > 0.05$	<1	0.954	1.572	−1.999	−1.381	−0.011	0.606
	Fusaric acid	$p > 0.05$	<1	0.250	0.175	0.264	0.189	−0.578	−0.653
Enniatins-Beauvericin toxins	Enniatin A	$p > 0.05$	<1	0.647	0.640	−0.418	−0.425	0.564	0.557
	Enniatin B1	$p > 0.05$	<1	−1.421	−0.724	0.767	1.464	−1.944	−1.248
	Enniatin B2	0.0434	1.039	0.058	0.436	−0.466	−0.088	0.458	0.836
	Beauvericin	0.0024	<1	0.438	0.656	−0.429	−0.211	−0.586	−0.368
Penicillium mycotoxins	Asperglaucide	$p > 0.05$	<1	−0.452	−0.802	−1.913	−2.263	1.347	0.997
	Pestalotin	$p > 0.05$	<1	0.088	0.167	−1.029	−0.951	0.981	1.0595
	Oxaline	$p > 0.05$	<1	0.428	0.475	−0.174	−0.127	−0.699	−0.652
	Flavoglaucin	0.0111	<1	−1.088	−0.992	0.842	0.938	−1.288	−1.192
	Cyclopenin	$p > 0.05$	<1	−0.515	−0.420	−0.281	−0.187	−0.393	−0.299
	Fellutanine A	$p > 0.05$	<1	1.067	1.129	0.037	0.099	−0.283	−0.221
Other fungal metabolites	Mycophenolic acid	0.0147	<1	−0.312	−0.094	−1.394	−1.175	1.296	1.515
	Ilicicolin A	0.0146	1.706	2.421	6.086	−4.303	−0.639	−2.547	1.117
	Ilicicolin B	0.0467	<1	−0.086	0.734	−3.767	−2.946	−1.029	−0.208
	Citreorosein	0.0369	<1	5.858	5.831	−2.225	−2.252	−2.510	−2.538
	Macrosporin	1.6×10^{-13}	<1	−4.451	−4.295	−0.415	−0.260	−4.296	−4.147
	Iso-Rhodoptilometrin	3.2×10^{-5}	1.254	−0.162	−0.462	−0.042	−0.341	−1.301	−1.599

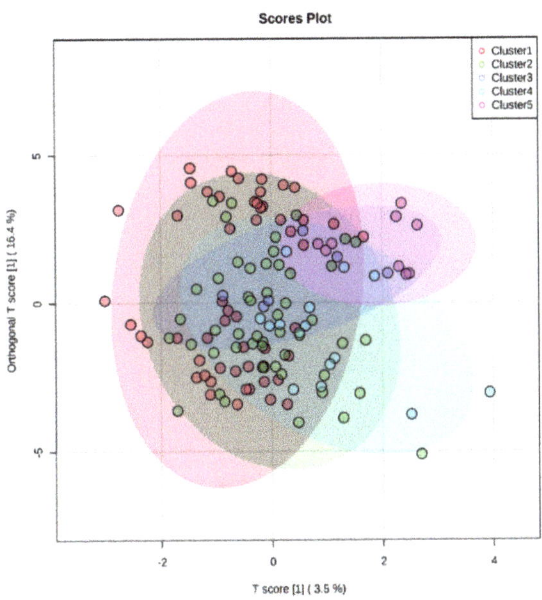

Figure 3. OPLS-DA score plot considering the different milk samples and their cluster, according to the corn silage contamination profile.

3.3. Discrimination of Milk Samples According to a k-Means Clustering Approach

The mycotoxin profile of the different milk samples showed a scarce prediction ability when considering as class discrimination parameter the cluster type used to previously classify the different corn silages [5]. This aspect is not surprising, considering that in our experimental conditions, several variables related to both animals and dairy farm conditions (including the other ingredients of the TMR) may have contributed to the profile observed, not only the contamination of corn silages by mycotoxins. However, the unsupervised statistical approach based on hierarchical clustering dendrogram (Table S1) revealed a tendency in the dataset to discriminate two main groups. This was confirmed by the PCA score plot resulting from the unsupervised k-means clustering approach (Figure 4), with two principal components able to explain the 36.8% of the total variability. Therefore, to find potential marker compounds of the discrimination observed, we used again a supervised statistical approach (OPLS-DA) combined with the S-plot to extrapolate the discriminant features. To this aim, two main groups (Group 1 and Group 2) were considered, and milk samples were assigned to these new groups, accordingly.

Afterwards, a new OPLS-DA prediction model considering the two new groups outlined by the unsupervised k-means clustering was built, and the resulting score plot is provided in Figure 5 (Figure 5A). The new OPLS-DA model was characterized by excellent parameters, recording a goodness-of-fit of 0.976 and a goodness-of-prediction of 0.965, recording a clear separation between the two groups of milk samples.

Finally, the S-plot related to the OPLS-DA score plot (Figure 5B) was inspected to extrapolate the most discriminant mycotoxins allowing the separation of both principal groups. Interestingly, we found that milk samples belonging to Group 2 were higher in apicidin and tentoxin, while those belonging to Group 1 were mainly discriminated by α-zearalenol and mycophenolic acid (Figure 5B). Considering the clear separation revealed by the OPLS-DA approach and based on the new sample grouping, we aim in a future work to explore more variables able to affect the final mycotoxin profile of milk. In fact, as widely reported in literature [8], various mycotoxins can modify the rumen flora due to their antimicrobial activity. This may decrease the degrading capacity of the rumen, resulting in an unexpected passage rate of intact toxins from other sources. Additionally, changes in the blood–milk barrier due to systemic, and particularly local, infections (mastitis) can affect the integrity of the blood–milk barrier and the pH gradient between blood and milk. Taken together, these effects could, in turn, alter the rate of excretion and facilitate the excretion of mycotoxins that are not expected in milk.

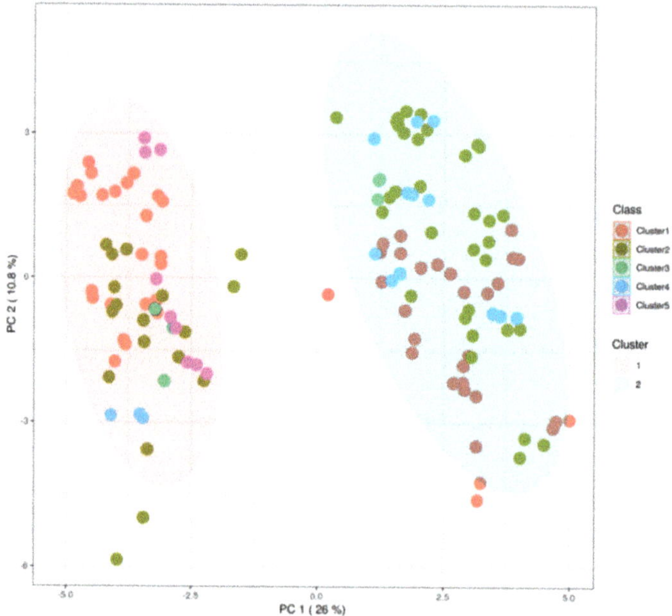

Figure 4. PCA score plot resulting from the unsupervised k-means clustering of the different milk samples under investigation, revealing two major groups according to the mycotoxin contamination profile.

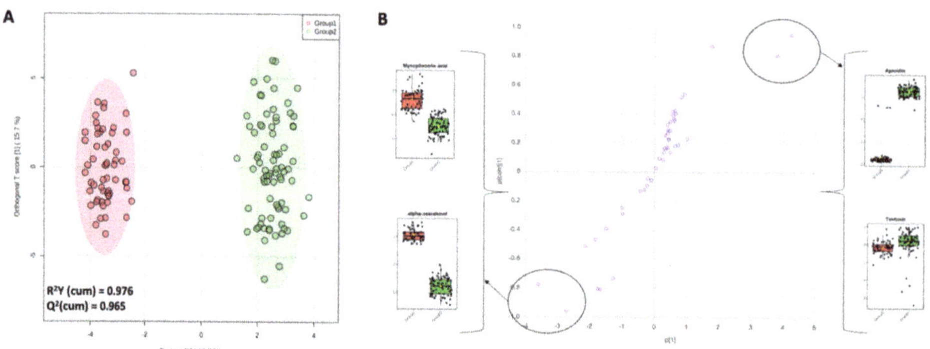

Figure 5. OPLS-DA score plot (**A**) considering the two group of milk samples outlined by the unsupervised k-means clustering approach, together with the S-plot; (**B**) outlining the most discriminant compounds for the comparison Group 1 vs. Group 2, namely mycophenolic acid and α-zearalenol (Group 1) followed by apicidin and tentoxin (Group 2).

3.4. Semi-Quantitative Analysis by UHPLC-HRMS of the Discriminant Markers

As a final step, a semi-quantification in HRMS-mode using authentic standards was carried out considering some of the discriminant marker compounds revealed by the different statistical approaches, namely α-zearalenol, mycophenolic acid, apicidin, and tentoxin. The results obtained are reported in Table 2, as average contents ($n = 3$) for each milk sample.

Table 2. Semiquantitative analysis based on HRMS (UHPLC-Orbitrap-MS) of tentoxin, α-zearalenol, mycophenolic acid, and apicidin in the different milk samples according to the main groups highlighted by the k-means clustering approach. nd = not detected.

Group (k-Means)	Milk Sample	Tentoxin (ng/mL)	α-Zearalenol (ng/mL)	Mycophenolic Acid (ng/mL)	Apicidin (ng/mL)
Group 1 (left side)	Sample 2	0.86	2.44	1.27	nd
	Sample 6	1.68	3.81	1.08	nd
	Sample 11	1.71	2.91	0.95	nd
	Sample 14	1.03	2.31	0.31	nd
	Sample 16	0.64	1.79	0.52	nd
	Sample 20	2.21	3.06	1.19	nd
	Sample 43	1.08	2.33	0.35	nd
	Sample 45	2.04	3.35	0.34	nd
	Sample 10	1.69	4.94	0.48	nd
	Sample 13	1.93	2.79	0.49	nd
	Sample 21	1.18	5.25	2.68	nd
	Sample 23	0.67	1.06	0.46	nd
	Sample 26	0.71	2.13	0.50	nd
	Sample 32	1.16	2.90	0.79	nd
	Sample 5	1.07	2.90	0.30	nd
	Sample 27	1.07	1.96	0.62	nd
	Sample 4	0.76	1.75	0.28	nd
	Sample 28	1.22	4.93	1.38	<0.1
	Sample 41	0.87	1.80	0.99	nd
Group 2 (right side)	Sample 19	1.01	<0.1	<0.1	0.33
	Sample 22	1.16	<0.1	<0.1	0.23
	Sample 25	2.35	<0.1	0.19	0.28
	Sample 29	1.85	<0.1	<0.1	0.15
	Sample 33	1.05	<0.1	0.16	0.11
	Sample 36	1.79	<0.1	<0.1	0.57
	Sample 37	2.07	<0.1	<0.1	0.27
	Sample 38	2.35	<0.1	<0.1	0.24
	Sample 39	1.16	<0.1	0.15	0.32
	Sample 40	2.06	<0.1	<0.1	<0.1
	Sample 1	0.29	<0.1	<0.1	0.13
	Sample 3	0.19	<0.1	<0.1	0.24
	Sample 7	1.49	<0.1	<0.1	0.20
	Sample 15	1.15	<0.1	<0.1	0.44
	Sample 17	1.28	<0.1	<0.1	<0.1
	Sample 24	1.32	<0.1	<0.1	0.43
	Sample 30	1.71	<0.1	<0.1	0.18
	Sample 31	2.42	<0.1	0.15	0.41
	Sample 34	2.02	<0.1	<0.1	0.55
	Sample 42	2.23	<0.1	<0.1	0.14
	Sample 44	2.88	<0.1	<0.1	0.41
	Sample 8	1.43	<0.1	<0.1	0.71
	Sample 9	<0.1	<0.1	<0.1	0.30
	Sample 12	1.01	<0.1	<0.1	0.15
	Sample 18	2.06	<0.1	<0.1	0.12
	Sample 46	1.46	<0.1	<0.1	0.25

As shown in the Table 2, we found a concentration range for the tentoxin of 0.19 to 2.9 ng/mL, with no significant differences between the different milk groups considered according to the k-means unsupervised clustering (i.e., Group 1 and Group 2). Regarding the distribution of tentoxin in the corn silages, Gallo et al. [5] reported a maximum content of 88.9 µg/kg for corn silage samples belonging to cluster 4, followed by cluster 2 (46.2 µg/kg) and cluster 5 (34.3 µg/kg). *Alternaria* species can produce more than 70 toxins, which play important roles in fungal pathogenicity and food safety, since some of them are harmful to humans and animals [25]. The studied *Alternaria* secondary metabolites belong to diverse chemical groups such as nitrogen-containing compounds (amide, cyclopeptides, etc.), steroids, terpenoids, pyranones, quinines, and phenolics. The major *Alternaria* toxins belong to the chemical groups dibenzo-pyrones, which include alternariol and alternariol monomethyl ether and cyclic tetrapeptides represented by tentoxin. These mycotoxins were the most studied metabolites produced by *Alternaria* strains on different substrates (tomato, wheat, blueberries, walnuts, etc.) and some of the main *Alternaria* compounds thought to pose a risk to human and animal health because of their known toxicity and their frequent presence as natural contaminants in food [26]. The cyclic tetrapeptide tentoxin is one of the major *Alternaria* toxins produced, along with dihydrotentoxin and isotentoxin. Their structures differ at the unsaturated bond of the N-methyldehydrophenylalanine moiety, which is hydrogenated into a single bond in dihydrotentoxin, and E configured in isotentoxin. All three compounds are phytotoxins, with tentoxin being the most potent, inhibiting photophosphorylation and inducing chlorosis [27]. However, no toxicological data are available for mammals, and the data on the occurrence of this toxin in food and feed are limited as well. Additionally, according to the scientific opinion provided by EFSA in 2016 [26], the levels of tentoxin in the different food categories considered were the lowest among the four *Alternaria* toxins covered. In this regard, the highest levels were found in samples of sunflower seeds (on average 80 µg/kg). Additionally, although based on limited data, vegetarians seemed to have higher dietary exposure to tentoxin than the general population. Overall, few data are available in literature about absolute quantification or screening of *Alternaria* toxins, such as tentoxin, in milk and dairy products. In a previous work, Izzo and co-authors [12] tentatively identified tentoxin in seven milk samples by using a high-resolution retrospective screening, pointing out the necessity of evaluating other fungal toxic metabolites in milk monitoring studies besides the regulated mycotoxins and their known metabolites.

Among the VIP markers of the OPLS-DA model we found also zearalenone (VIP score = 1.358) and its metabolite α-zearalenol (VIP score = 1.098). As reported in Gallo et al. [5], the corn silages belonging to cluster 1 were characterized by a total content of zearalenone and derivatives of 95.3 µg/kg, with a great abundance of zearalenone-sulfone (i.e., 91.2 µg/kg). On the other hand, the lowest contents for zearalenone derivatives were reported for cluster 2 (22 µg/kg) and cluster 4 (2.5 µg/kg), whilst the highest values were reported for cluster 3 (i.e., 152.8 µg/kg) and cluster 5 (i.e., 286.6 µg/kg). Accordingly, the highest abundance of zearalenone-sulfone was found in cluster 5, being 284.9 µg/kg. This might be of great concern considering the toxic effects of this compound. Zearalenone and its derivative, α-zearalenol, are a family of phenolic compounds produced by several species of *Fusarium* (such as *F. graminearum*, *F. culmorum*, *F. crookwellense*, *F. sambucinum* and *F. equiseti*), which can infect many important crops such as corn, wheat, sorghum, barley, oats, sesame seed, hay, and corn silage [28]. These fungal toxins have been associated with hyperestrogenism and other reproductive disorders in swine. In sows, a series of reproductive disorders may occur at greater levels of zearalenone in feed (50–100 mg/g feed), including the induction of vulvo-vaginitis, vaginal and rectal prolapses, delayed onset of the first estrus, infertility characterized by continuous estrus, pseudo-pregnancy, ovarian abnormalities, and pregnancy loss [29]. Additionally, the α-zearalenol metabolite is reported to be three/four times more estrogenic than zearalenone [30]. In our experimental conditions, we found not quantifiable values up to more than 5 ng/mL for α-zearalenol. Interestingly, milk samples belonging to cluster 5 showed the higher average value, according to the highest values of zearalenone derivatives recorded in the silages belonging to cluster 5 (i.e., 286.6 µg/kg). Besides, α-zearalenol was a specific marker of Group 1 of the OPLS-DA model reported in Figure 5, while Group 2 showed no detectable levels (Table 2).

Finally, mycophenolic acid (a *Penicillium* mycotoxin) and apicidin (a *Fusarium* mycotoxin) were found to be two of the most discriminant mycotoxins of the new OPLS-DA model built (Figure 5). To date, no comprehensive information concerning the carry-over of mycophenolic acid from feed to milk is available, although carry-over is possible [31]. However, in our experimental conditions, we found semi-quantitative values of <0.1 up to 1.38 ng/mL in milk samples for this *Penicillium* mycotoxin. This compound presents low acute cytotoxicity on the human intestinal cell line Caco-2 compared to other mycotoxins (such as T-2 toxin, gliotoxin, deoxynivalenol, and patulin) after 48 h exposure; however, it has been shown to possess immunosuppressive effects [31]. According to literature, mycophenolic acid is found mainly in blue-veined cheeses. In fact, Fontaine et al. [31] reported that the 75% of cheese samples analyzed contained a maximum of 705 µg/kg of mycophenolic acid. However, cheeses may be directly contaminated by mycotoxins, such as mycophenolic acid, because of accidental or intentional mycotoxigenic fungal development on the cheese surface or in their core, while the presence of mycophenolic acid in bulk milk is only attributable to potential carry-over from the feed. Therefore, further studies based on strong and targeted monitoring plans of mycophenolic acid in both the TMR and milk (considering individual animals and single diets) are mandatory. As revealed by the semi-quantification of mycophenolic acid reported in Table 2, this compound was a specific marker of the Group 1. Regarding apicidin, no information in the literature is available about its presence in milk and dairy products, and this was also due to its nature of "emerging" mycotoxin. Overall, apicidin (a cyclic tetrapeptide) is a fungal metabolite that exhibits potent, broad-spectrum, antiprotozoal activity and inhibits histone deacetylase activity at nanomolar concentrations [32]. As reported in Table 2, this compound was a specific marker of the Group 2. Apicidin and mycophenolic acid (as emerging mycotoxins) have been identified also in the corn silages previously analyzed by Gallo et al. [5]; however, in our experimental conditions it is impossible to provide carry-over evaluations and further work is mandatory to better evaluate the impact and incidence of the contamination profile.

4. Conclusions

In this work, we have assessed the potential of a metabolomics-based approach coupled with a retrospective screening by high-resolution mass spectrometry to evaluate the mycotoxin profile (in terms of both regulated and emerging mycotoxins) of bulk milk samples. Overall, the HRMS approach allowed us to identify 46 mycotoxins, but the accuracy of the prediction was not robust enough and further ad hoc and targeted studies (based on dedicated and optimized extraction conditions and tandem MS analyses) are required to confirm the robustness of the markers proposed. Additionally, by coupling unsupervised (different clustering algorithms) and supervised (such as OPLS-DA) approaches we found the most discriminant mycotoxins driving the separations of the bulk milk samples under investigation. In this regard, among the most discriminant markers we found α-zearalenol, mycophenolic acid, tentoxin, and apicidin. The preliminary semi-quantitative results obtained in this work suggest potential carry-over and metabolization phenomena in milk of the selected mycotoxins, although further ad hoc confirmation studies are mandatory, mainly when considering the toxicity reported in literature for some of the markers identified.

Supplementary Materials: The following are available online at https://www.mdpi.com/article/10.3390/foods10092025/s1, Table S1: dataset resulting from the screening by HRMS of the different mycotoxins annotated in the milk samples (the MS1 and MS/MS isotopic profiles for each compound are provided with both isotopic mass and corresponding relative intensity values) together with the sample legend and the unsupervised grouping (tree plot) of the different milk samples according to the hierarchical clustering approach (built considering the Euclidean similarity measure and 'Wards' as linkage rule).

Author Contributions: Conceptualization, A.G.; Methodology, G.R.; Software, G.R.; Validation, G.R.; Formal analysis, G.R. and F.G.; Investigation, G.R., F.G. and A.G.; Data curation, G.R. and A.G.; Writing—original draft preparation, G.R. and A.G.; Writing—review and editing, F.M. and A.G.; Visualization, G.R. and F.G.; Supervision, A.G. and F.M. All authors have read and agreed to the published version of the manuscript.

Funding: This research received no external funding.

Data Availability Statement: Not applicable.

Acknowledgments: This work was supported by the Cremona Agri-food technologies project (CRAFT) funded by Fondazione Cariplo and Regione Lombardia.

Conflicts of Interest: The authors declare no conflict of interest.

References

1. Zhu, D.; Kebede, B.; McComb, K.; Hayman, A.; Chen, G.; Frew, R. Milk Biomarkers in Relation to Inherent and External Factors Based on Metabolomics. *Trends Food. Sci. Technol.* **2021**, *109*, 51–64. [CrossRef]
2. Rocchetti, G.; O'Callaghan, T.F. Application of Metabolomics to Assess Milk Quality and Traceability. *Curr. Opin. Food Sci.* **2021**, *40*, 168–178. [CrossRef]
3. Joubran, A.M.; Pierce, K.M.; Garvey, N.; Shalloo, L.; O'Callaghan, T.F. Invited Review: A 2020 Perspective on Pasture-Based Dairy Systems and Products. *J. Dairy Sci.* **2021**, *104*, 7364–7382. [CrossRef]
4. Gallo, A.; Fancello, F.; Ghilardelli, F.; Zara, S.; Froldi, F.; Spanghero, M. Effects of Several Lactic Acid Bacteria Inoculants on Fermentation and Mycotoxins in Corn Silage. *Anim. Feed Sci. Technol.* **2021**, *277*, 114962. [CrossRef]
5. Gallo, A.; Ghilardelli, F.; Atzori, A.S.; Zara, S.; Novak, B.; Faas, J.; Fancello, F. Co-Occurrence of Regulated and Emerging Mycotoxins in Corn Silage: Relationships with Fermentation Quality and Bacterial Communities. *Toxins* **2021**, *13*, 232. [CrossRef] [PubMed]
6. Gallo, A.; Giuberti, G.; Frisvad, J.C.; Bertuzzi, T.; Nielsen, K.F. Review on Mycotoxin Issues in Ruminants: Occurrence in Forages, Effects of Mycotoxin Ingestion on Health Status and Animal Performance and Practical Strategies to Counteract Their Negative Effects. *Toxins* **2015**, *7*, 3057–3111. [CrossRef]
7. Liew, W.-P.-P.; Mohd-Redzwan, S. Mycotoxin: Its Impact on Gut Health and Microbiota. *Front. Cell. Infect. Microbiol.* **2018**, *8*, 60. [CrossRef] [PubMed]
8. Fink-Gremmels, J. Mycotoxins in Cattle Feeds and Carry-Over to Dairy Milk: A Review. *Food Addit. Contam. Part A* **2008**, *25*, 172–180. [CrossRef] [PubMed]
9. Kemboi, D.C.; Antonissen, G.; Ochieng, P.E.; Croubels, S.; Okoth, S.; Kangethe, E.K.; Faas, J.; Lindahl, J.F.; Gathumbi, J.K. A Review of the Impact of Mycotoxins on Dairy Cattle Health: Challenges for Food Safety and Dairy Production in Sub-Saharan Africa. *Toxins* **2020**, *12*, 222. [CrossRef] [PubMed]
10. Ogunade, I.M.; Martinez-Tuppia, C.; Queiroz, O.C.M.; Jiang, Y.; Drouin, P.; Wu, F.; Vyas, D.; Adesogan, A.T. Silage Review: Mycotoxins in Silage: Occurrence, Effects, Prevention, and Mitigation. *J. Dairy Sci.* **2018**, *101*, 4034–4059. [CrossRef] [PubMed]
11. Huang, L.C.; Zheng, N.; Zheng, B.Q.; Wen, F.; Cheng, J.B.; Han, R.W.; Xu, X.M.; Li, S.L.; Wang, J.Q. Simultaneous Determination of Aflatoxin M1, Ochratoxin A, Zearalenone and α-zearalenol in Milk by UHPLC–MS/MS. *Food Chem.* **2014**, *146*, 242–249. [CrossRef]
12. Izzo, L.; Rodríguez-Carrasco, Y.; Tolosa, J.; Graziani, G.; Gaspari, A.; Ritieni, A. Target Analysis and Retrospective Screening of Mycotoxins and Pharmacologically Active Substances in Milk Using an Ultra-High-Performance Liquid Chromatography/High-Resolution Mass Spectrometry Approach. *J. Dairy Sci.* **2020**, *103*, 1250–1260. [CrossRef]
13. Tamura, M.; Mochizuki, N.; Nagatomi, Y.; Harayama, K.; Toriba, A.; Hayakawa, K. A Method for Simultaneous Determination of 20 *Fusarium* Toxins in Cereals by High-Resolution Liquid Chromatography-Orbitrap Mass Spectrometry with a Pentafluorophenyl Column. *Toxins* **2015**, *7*, 1664–1682. [CrossRef] [PubMed]
14. Castaldo, L.; Graziani, G.; Gaspari, A.; Izzo, L.; Tolosa, J.; Rodríguez-Carrasco, Y.; Ritieni, A. Target Analysis and Retrospective Screening of Multiple Mycotoxins in Pet Food Using UHPLC-Q-Orbitrap HRMS. *Toxins* **2019**, *11*, 434. [CrossRef] [PubMed]
15. De Dominicis, E.; Commissati, I.; Gritti, E.; Catellani, D.; Suman, M. Quantitative Targeted and Retrospective Data Analysis of Relevant Pesticides, Antibiotics and Mycotoxins in Bakery Products by Liquid Chromatography-Single-Stage Orbitrap Mass Spectrometry. *Food Addit. Contam. Part A* **2015**, *32*, 1617–1627. [CrossRef] [PubMed]
16. Rocchetti, G.; Ghilardelli, F.; Bonini, P.; Lucini, L.; Masoero, F.; Gallo, A. Changes of Milk Metabolomic Profiles Resulting from a Mycotoxins-Contaminated Corn Silage Intake by Dairy Cows. *Metabolites* **2021**, *11*, 475. [CrossRef]
17. Rocchetti, G.; Gallo, A.; Nocetti, M.; Lucini, L.; Masoero, F. Milk Metabolomics Based on Ultra-High-Performance Liquid Chromatography Coupled with Quadrupole Time-of-Flight Mass Spectrometry to Discriminate Different Cows Feeding Regimens. *Food Res. Int.* **2020**, *134*, 109279. [CrossRef]
18. Bellassi, P.; Rocchetti, G.; Nocetti, M.; Lucini, L.; Masoero, F.; Morelli, L. A Combined Metabolomic and Metagenomic Approach to Discriminate Raw Milk for the Production of Hard Cheese. *Foods* **2021**, *10*, 109. [CrossRef]
19. Tsugawa, H.; Cajka, T.; Kind, T.; Ma, Y.; Higgins, B.; Ikeda, K.; Kanazawa, M.; Vander Gheynst, J.; Fiehn, O.; Arita, M. MS-DIAL: Data-Independent MS/MS Deconvolution for Comprehensive Metabolome Analysis. *Nat. Methods.* **2015**, *12*, 523–526. [CrossRef] [PubMed]
20. Tsugawa, H.; Kind, T.; Nakabayashi, R.; Yukihira, D.; Tanaka, W.; Cajka, T.; Saito, K.; Fiehn, O.; Arita, M. Hydrogen Rearrangement Rules: Computational MS/MS Fragmentation and Structure Elucidation Using MS-FINDER Software. *Anal. Chem.* **2016**, *88*, 7946–7958. [CrossRef]
21. Salek, R.M.; Steinbeck, C.; Viant, M.R.; Goodacre, R.; Dunn, W.B. The Role of Reporting Standards for Metabolite Annotation and Identification in Metabolomic Studies. *GigaScience* **2013**, *2*, 2047-217X. [CrossRef] [PubMed]

22. Zachariasova, M.; Cajka, T.; Godula, M.; Malachova, A.; Veprikova, Z.; Hajslova, J. Analysis of Multiple Mycotoxins in Beer Employing (Ultra)-High-Resolution Mass Spectrometry. *Rapid. Commun. Mass Spectrom.* **2010**, *24*, 3357–3367. [CrossRef] [PubMed]

23. Pang, Z.; Chong, J.; Zhou, G.; de Lima Morais, D.A.; Chang, L.; Barrette, M.; Gauthier, C.; Jacques, P.É.; Li, S.; Xia, J. MetaboAnalyst 5.0: Narrowing the Gap Between Raw Spectra and Functional Insights. *Nucleic Acids Res.* **2021**, *49*, W388–W396. [CrossRef] [PubMed]

24. Pang, Z.; Chong, J.; Li, S.; Xia, J. MetaboAnalystR 3.0: Toward an Optimized Workflow for Global Metabolomics. *Metabolites* **2020**, *10*, 186. [CrossRef]

25. Escrivá, L.; Oueslati, S.; Font, G.; Manyes, L. Alternaria Mycotoxins in Food and Feed: An Overview. *J. Food Qual.* **2017**, *2017*, 1569748. [CrossRef]

26. Arcella, D.; Eskola, M.; Gómez Ruiz, J.A. Dietary Exposure Assessment to *Alternaria* Toxins in the European Population. *EFSA J.* **2016**, *14*, e04654.

27. Gotthardt, M.; Asam, S.; Gunkel, K.; Moghaddam, A.F.; Baumann, E.; Kietz, R.; Rychlik, M. Quantitation of Six Alternaria Toxins in Infant Foods Applying Stable Isotope Labeled Standards. *Front. Microbiol.* **2019**, *10*, 109. [CrossRef] [PubMed]

28. Arie, T. Fusarium diseases of cultivated plants, control, diagnosis, and molecular and genetic studies. *J. Pest. Sci.* **2019**, *44*, 275–281. [CrossRef]

29. Minervini, F.; Dell'Aquila, M.E. Zearalenone and Reproductive Function in Farm Animals. *Int. J. Mol. Sci.* **2008**, *9*, 2570–2584. [CrossRef]

30. Keller, L.; Abrunhosa, L.; Keller, K.; Rosa, C.A.; Cavaglieri, L.; Venâncio, A. Zearalenone and Its Derivatives α-Zearalenol and β-Zearalenol Decontamination by *Saccharomyces cerevisiae* Strains Isolated from Bovine Forage. *Toxins* **2015**, *7*, 3297–3308. [CrossRef]

31. Fontaine, K.; Passeró, E.; Vallone, L.; Hymery, N.; Coton, M.; Jany, J.-L.; Mounier, J.; Coton, E. Occurrence of Roquefortine C, Mycophenolic Acid and Aflatoxin M1 Mycotoxins in Blue-Veined Cheeses. *Food Cont.* **2015**, *47*, 634–640. [CrossRef]

32. Martínez-Iglesias, O.; Cacabelos, R. Chapter 12—Epigenetic Treatment of Neurodegenerative disorders. *Histone Modif. Ther.* **2020**, *20*, 311–335.

Article

Small RNAs, Degradome, and Transcriptome Sequencing Provide Insights into Papaya Fruit Ripening Regulated by 1-MCP

Jiahui Cai, Ziling Wu, Yanwei Hao, Yuanlong Liu, Zunyang Song, Weixin Chen, Xueping Li and Xiaoyang Zhu *

Guangdong Provincial Key Laboratory of Postharvest Science of Fruits and Vegetables/Engineering Research Center for Postharvest Technology of Horticultural Crops in South China, Ministry of Education, College of Horticulture, South China Agricultural University, Guangzhou 510642, China; jh.cai_chn@outlook.com (J.C.); ziling_ng@163.com (Z.W.); yanweihao@scau.edu.cn (Y.H.); liuyuanlong@scau.edu.cn (Y.L.); songzunyang@163.com (Z.S.); wxchen@scau.edu.cn (W.C.); lxp88@scau.edu.cn (X.L.)
* Correspondence: xiaoyang_zhu@scau.edu.cn; Tel./Fax: +86-20-85288280

Abstract: As an inhibitor of ethylene receptors, 1-methylcyclopropene (1-MCP) can delay the ripening of papaya. However, improper 1-MCP treatment will cause a rubbery texture in papaya. Understanding of the underlying mechanism is still lacking. In the present work, a comparative sRNA analysis was conducted after different 1-MCP treatments and identified a total of 213 miRNAs, of which 44 were known miRNAs and 169 were novel miRNAs in papaya. Comprehensive functional enrichment analysis indicated that plant hormone signal pathways play an important role in fruit ripening. Through the comparative analysis of sRNAs and transcriptome sequencing, a total of 11 miRNAs and 12 target genes were associated with the ethylene and auxin signaling pathways. A total of 1741 target genes of miRNAs were identified by degradome sequencing, and nine miRNAs and eight miRNAs were differentially expressed under the ethylene and auxin signaling pathways, respectively. The network regulation diagram of miRNAs and target genes during fruit ripening was drawn. The expression of 11 miRNAs and 12 target genes was verified by RT-qPCR. The target gene verification showed that *cpa-miR390a* and *cpa-miR396* target *CpARF19-like* and *CpERF RAP2-12-like*, respectively, affecting the ethylene and auxin signaling pathways and, therefore, papaya ripening.

Keywords: 1-MCP treatments; microRNA; degradome and transcriptome; papaya ripening; ethylene and auxin

Citation: Cai, J.; Wu, Z.; Hao, Y.; Liu, Y.; Song, Z.; Chen, W.; Li, X.; Zhu, X. Small RNAs, Degradome, and Transcriptome Sequencing Provide Insights into Papaya Fruit Ripening Regulated by 1-MCP. *Foods* **2021**, *10*, 1643. https://doi.org/10.3390/foods10071643

Academic Editors: Yelko Rodríguez-Carrasco and Bojan Šarkanj

Received: 8 June 2021
Accepted: 13 July 2021
Published: 15 July 2021

Publisher's Note: MDPI stays neutral with regard to jurisdictional claims in published maps and institutional affiliations.

1. Introduction

Papaya (*Carica papaya* L.) is a popular and widely planted fruit in tropical and subtropical regions due to its unique taste, nutritional benefits, and medicinal benefits [1–3]. As one of the classic climacteric fruits, papaya fruits are highly perishable after harvest as a result of rapid ripening and softening and susceptibility to biotic or abiotic stresses [4,5], which usually result in a high percentage of product loss [6,7].

The ethylene antagonist 1-methylcyclopropene (1-MCP) has been widely used to maintain the quality and extend the shelf life of harvested products [8]. By binding to ethylene receptors, 1-MCP can inhibit the release of ethylene, inhibit the respiratory intensity of fruits and vegetables, and delay the ripening of various fruits and vegetables [9–12]. 1-MCP treatment on papaya at earlier development stages or for long-term/high-concentration treatments will lead to a "rubbery texture" phenomenon, where the fruits are unable to completely soften during later storage, leading to tasteless lack of flavor [13]. RNA-seq analysis has shown that improper 1-MCP treatment severely inhibits cell wall degradation and fruit softening by inhibiting ethylene signal transduction and cell wall metabolism pathways [10]. Integrated analysis of metabolomics and RNA-seq data has shown that various energy metabolites for the tricarboxylic acid cycle, glycolic acid cycle, flavonoids, and phenylpropane pathways were significantly affected by 1-MCP, all of which play important roles in fruit ripening and the softening disorder caused by improper 1-MCP treatment ($400 \cdot L^{-1} \cdot 16$ h) [14]. However, a deeper understanding of the ripening and softening disorder caused by inappropriate 1-MCP treatment is imperative for further commercial use of 1-MCP.

Ethylene and auxin play critical roles in fruit ripening [15–17]. At the transcript level, various transcription factors act as important regulators in the auxin and ethylene signaling pathways and regulate fruit ripening. In persimmon fruit, the *DkERF8/16/18* genes may participate in fruit ripening by accelerating cell wall modification and ethylene biosynthesis [18]. In papaya, it was found that CpNAC3 interacted with CpMADS4 and regulated the role of ethylene signal transcription factors, namely CpERF9 and CpEIL5, to regulate fruit ripening [19]. CpEBF1 interacts with CpMADS1 and regulates cell wall degradation-related genes to modulate the fruit ripening process and softening disorder caused by 1-MCP [13]. Exogenous auxin delays the ripening process of tomato fruits by inhibiting the production of ethylene, the accumulation of carotenoids, and the degradation of chlorophyll [20]. Auxin-induced *DzARF2A* expression was confirmed in response to exogenous auxin application, indicating the auxin-mediated *DzARF2A* role in durian fruit ripening [21]. Through transcriptome sequencing, it was found that most of the differentially expressed genes between suitable and improper 1-MCP treatment groups were enriched in starch and sucrose metabolism, carbon metabolism, plant hormone signal transduction, and amino acid biosynthesis pathways [10]. Among these differentially expressed genes, there were 21 genes enriched in ethylene and auxin signaling pathways [10].

MicroRNA (miRNA) is a type of non-coding small RNA that is around 21 nucleotide (nt). They are widely presented across eukaryotes, where most are negative regulators of gene expression [22,23]. They mainly regulate the expression of plant genes at the post-transcriptional level by mediating cleavage of mRNA target molecules or reducing the translation of target molecules, thereby regulating morphogenesis, growth, development, plant organ hormone secretion, plant organ signal transduction, and the ability to respond to stressors in the environment [24–26]. Recently, a growing body of studies have found that small RNAs were involved in regulating auxin and ethylene signal transduction. Li et al. [27] found that *miR160*, *miR390*, and their target genes were related to auxin signaling and participate in the pathway of adventitious root formation in "M9-T337" apple rootstock. Auxin response factor (ARF) is a plant-specific transcription factor that mediates the downstream expression of auxin response genes by binding to auxin response elements and participating in various processes during plant growth and development [28]. The gene families *TaARF1*, *TaARF4*, *TaARF7*, *TaARF34*, and *TaARF39*, located in the wheat genome, were predicted to be targets of Tae-miRNA160, a stress-responsive miRNA. This showed that they were under the regulation of early auxin-responsive gene expression in the auxin signaling pathway [29]. An analysis of the differential expression of miRNAs in banana under ethylene treatment found that *miR162a*, *miR167a*, *miR172a*, and *miR319a* participated in ethylene-dependent fruit ripening [30]. All these results suggested that miRNAs play critical roles in plant development and fruit ripening by regulating ethylene targets, auxin, and other signaling pathways.

This study integrates a differential expression analysis of miRNA, RNA-seq, and degradation sequencing in papaya fruit under different 1-MCP treatments in an effort to identify the key miRNAs and their targets that regulate papaya fruit ripening and the ripening disorder. This work further explores the transcriptional regulation network in papaya fruit and provides new ideas for future physiological and molecular research.

2. Materials and Methods

2.1. Plant Material and Treatment

Papaya (*Carica papaya* L. cv. 'suiyou-2') fruits were harvested from a local commercial farm in the Panyu district of Guangzhou, Guangdong, China. The fruit were harvested at the color-breaking stage of maturity (5% yellow < peel color < 15% yellow) as long as they were disease-free, packaged into polystyrene boxes, and then transported immediately to the laboratory. The harvested fruit were dipped in 0.2% (*w/v*) hypochlorite solution for 10 min and then soaked in a 500 mg·mL^{-1} solution of mixed prochloraz (Huifeng, Jiangsu, China) and iprodione (Kuaida, Jiangsu, China) for 1 min to minimize the effect of microbes/microbe contamination. Treatments were then carried out with the same protocol as in our previous work in Zhu et al. [31]. A total of 600 papaya fruits were selected and divided into three sub-groups, and each group contained 200 papaya fruits. Two groups of the fruit were treated with 400 nL·L^{-1} 1-MCP in a closed foam box for 2 and 16 h, respectively; air treatment for 16 h was used as a control. All fruits were then treated with 1000 µL·L^{-1} ethephon for 1 min. After air-drying for a few minutes, the papaya fruit was packed in an unsealed polyethylene bag (10 cm × 20 cm, 0.02 mm thick) and stored at 22 °C. All the treatments were conducted with three biological replicates.

Three sampling points at days 0, 1, and 6 were selected for measurement with three biological replicates. For each sampling point, three fruits representing the three biological replicates were collected. Samples were taken from the middle of the papaya fruit flesh, frozen immediately using liquid nitrogen, and then stored at −80 °C for further testing. Subsequent usage included RNA extraction, small RNA sequencing, and degradome sequencing.

2.2. RNA Extraction and cDNA Library Preparation

For RNA-seq, all RNA extraction and library preparation procedures were conducted as described in Zhu et al. [31].

The NEB Next Ultra-small RNA Library Preparation Kit for Illumina (NEB, #E7530, Ipswich, MA, USA) was used according to the manufacturer's protocol in order to generate a small RNA-seq library. Library quality was evaluated using an Agilent Bioanalyzer 2100 system. Samples from the control group, short-term 1-MCP treatment group, and long-term 1-MCP treatment group were selected for miRNA analysis after storage for 0, 1, and 6 days. Each sample time contained three biological replicates, and a total of 21 libraries were constructed. After removing the adapter sequences and low-quality sequence reads (including reads containing more than 10% N, reads without 3′ linker sequences, and sequences shorter than 18 nucleotides or longer than 30 nucleotides), the clean reads were then mapped to the papaya reference genome (http://www.plantgdb.org/CpGDB/). Using a BLASTN search against miRbase (V21), known miRNAs were identified from mapped small RNA tags. For the new miRNA candidates, we used miRDeep2 for their identification. The total number of identified miRNAs in each constructed library was then normalized to TPM (number of transcripts per million of the clean tags). The DEGseq R package was used for differential expression analysis between the different groups. The q-value was used to adjust the *p*-value. Small RNAs with | log2(foldchange) | > 1 and *p* -value < 0.01 were assigned as differentially expressed. Target Finder software was used to predict the target genes of differentially expressed miRNAs (http://targetfinder.org/). The target gene identification was also based on the integration of small RNAs and transcriptome sequencing analysis. Normally, miRNA and mRNA pairs have a negative regulatory network relationship regarding their expression. GO software (http://www.geneontology.org/) was used to enrich and analyze the differential target genes between sample groups, and the KEGG (Kyoto Encyclopedia of Genes and Genomes) database was used to analyze the pathway annotation of differentially expressed miRNA target genes (http://www.genome.jp/kegg/).

The papaya samples, including fruits treated with 400 nL· L^{-1} 1-MCP for 2 or 16 h and from the control group, were mixed and used to establish a degradation group library. After the samples were extracted, total RNA was extracted using the reagent Trizol (Invitrogen, CA, USA). The quantity and purity of the total RNA were analyzed using a Bioanalyzer 2100 and RNA 6000 Nano Lab Chip Kit (Agilent, CA, USA) with an RIN number > 7.0. Approximately 20 µg of total RNA was used for degradome library construction. Single-end sequencing (36 bp) was performed on an Illumina Hiseq2500 at LC-BIO (Hangzhou, China) according to the protocol recommended by the sequencing facility. Raw sequencing reads containing no adaptors nor low-quality reads were obtained using Illumina's software. The filtered sequencing reads were then used to identify potentially cleaved targets using the CleaveL pipeline. Degradome data and reads were mapped to the mRNA downloaded from JGI (http://www.plantgdb.org/CpGDB/). Only perfectly matched alignments to the given read were preserved for further degradation analysis.

2.3. RT-qPCR Verification

In order to verify the expression profiles of the identified miRNA and its target mRNA, 7 known miRNAs, 4 novel miRNAs, and 12 corresponding target genes were selected for real-time quantitative PCR (RT-qPCR) validation.

Total RNA and small RNAs were extracted using TRIzol reagent and Fruit-mate for RNA purification (Takara, Japan). Total RNA was reverse-transcribed using stem-loop qRT-PCR [32]. Stem-loop primers for reverse transcription and primers for RT-qPCR are listed in Tables S1 and S2. The relative gene expression value was normalized against the relative value of the *CpTBP1* gene for mRNA expression [33] and 5 s RNA for miRNA expression [34]. SuperScript III reverse transcriptase (Invitrogen, Carlsbad, California, USA) was used to reverse transcribe the total RNA using the pulse reverse transcription program. The RT-qPCRs were performed using a total volume of 20 µL containing 10 µL SYBR Mixture (Promega, Madison, Wisconsin, USA), 3 µL cDNA template, 0.5 mM primers, and 6 µL ddH2O. PCR was performed on a Bio-Rad CFX96 real-time PCR system using the qPCR Master Mix Kit (Promega, Madison, WI, USA). Three biological replicates were used to determine the expression of each gene, and expression was calculated using the $2^{-\Delta\Delta CT}$ method [33].

2.4. Plant Expression Vector Construction and Transformation

Validated miRNAs were cloned into a pBI121 binary vector driven by a cauliflower mosaic virus 35S promoter (CaMV35S). The miRNAs' corresponding target genes were cloned into a pMS4 vector containing green fluorescent protein (GFP). Vector constructions were conducted in reference to Wang's method [35]. Confirmation of the target gene site was identified using the website http://plantgrn.noble.org/psRNATarget/). The primers are listed in Table S3. The constructed vectors were then transformed into tobacco mediated by *Agrobacterium tumefaciens* (GV3101). Young tobacco leaves were cultivated for five weeks and used for injection of *Agrobacterium tumefaciens*. Transformed leaves were photographed with ultraviolet light after 36 h of infection.

2.5. Statistical Analysis

Each treatment was conducted in three independent biological replicates, and the variance of the collected data was analyzed. Statistical differences between groups were assessed based on Duncan's Multiple Range Test (DMRT) in SPSS 19.0 (IBM, Armonk, NY, USA). The graphics were drawn using Prism 8 software (GraphPad Inc., La Jolla, CA, USA) and Adobe Illustrator software (CC 2017; Adobe Inc., San Jose, CA, USA).

3. Results

3.1. Construction and Sequencing of Small RNA Libraries

Our previous work showed that 1-MCP treatment (400 nL·L^{-1}, 2 h) can delay the ripening of papaya fruit, but improper 1-MCP treatment (400 nL·L^{-1}, 16 h) causes an adverse "rubbery" texture [31]. Therefore, small RNA (sRNA) sequencing was performed. A total of 354.06 megabytes (MB) of clean reads was obtained from 21 different libraries. For each sample, low-quality sequences and reads with an unknown base (N) greater than 10% of the sequence were first removed and then 3$'$ adaptor sequences, reads with low complexity, and reads homologous to t/rRNA sequence were removed. The data for each sample consisted of no less than 12.23 MB of clean reads (Table S4). In addition, 98.77 to 99.11% of the total reads ranging from 18 to 30 nt library lengths exactly matched the papaya genome sequence. The length distribution of the sRNAs was mainly concentrated between

18 and 24 nt, where the papaya libraries were mainly composed of 20 and 21 nt length sRNAs (Figure 1A). However, in the papaya injected with papaya ringspot virus (PRSV), 21 and 24 nt length sRNAs accounted for the majority of the total sRNAs present [36]. One possible reason for the observed difference in mature sRNA lengths across the different groups may be due to the differential activity across the various sRNA biogenesis pathways. Nucleotide sequence analysis of the identified miRNAs showed that uridine (U) was the most common nucleotide in sRNAs of 20–24 nt length, while cytosine (C) had a higher enrichment in the sRNAs of 19 and 25 nt lengths (Figure 1B). C and U account for the majority of the total number of bases in papaya sRNA.

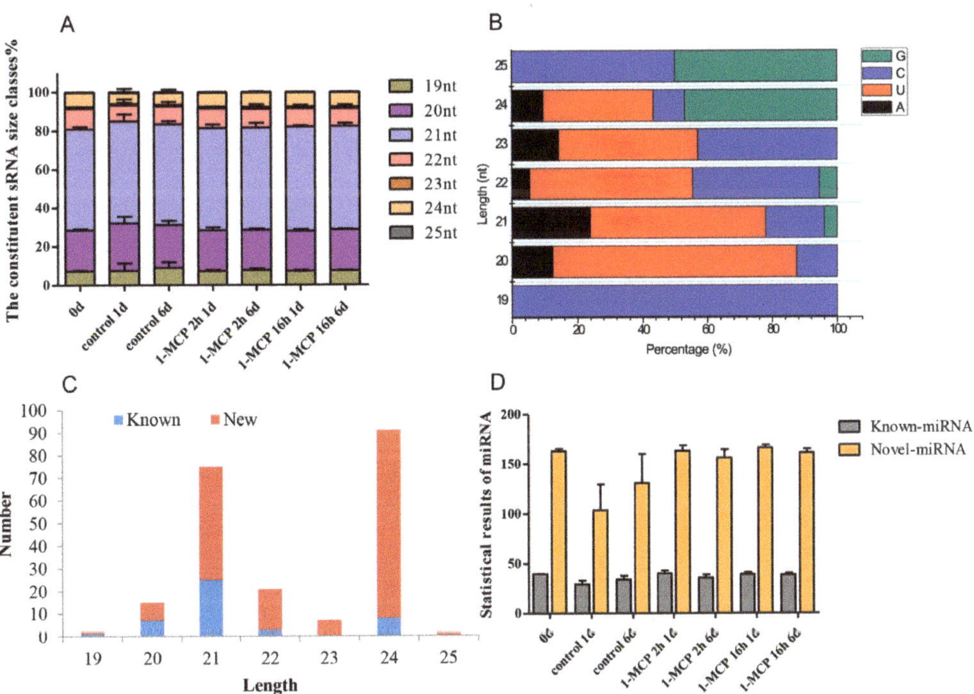

Figure 1. Changes in the distribution and complexity of sRNA in papaya samples. (**A**) Sizes of the 19–25 nt sRNA readings are classified and distributed under different 1-MCP treatments. The composition of sRNAs at each storage time point is shown, defined as 0 d, 1-MCP 2 h 1 d and 6 d, and 1-MCP 16 h 1 d and 6 d. The rRNA and tRNA sequences were deleted. (**B**) Base preference distribution of different sRNA reads. (**C**) sRNA read length classification distribution of known and novel miRNAs. (**D**) Statistical results of miRNA for each sample, including known and novel miRNAs.

3.2. Identification of Known and Novel miRNAs

Known miRNAs in papaya fruit were identified through sequence matching with an miRNA database (miRBase v21). Read sequences that were exactly the same as the known miRNA were considered to be an identified known miRNA. A total of 213 miRNAs were identified in the present study, 44 of which are already known miRNAs, and 169 are newly predicted miRNAs (Table S5). As shown in Figure 1C, the distribution of known and new miRNAs in sRNAs is different across different lengths. The number of new miRNAs discovered was significantly greater than the number of known miRNAs. Among the newly discovered miRNAs, the number of 21 and 24 nt length miRNAs was significantly greater than that of those of other size classes. The distribution of known miRNAs and new miRNAs from each sample is shown in Figure 1D. Fewer miRNAs were identified in the 1 day (s) after treatment (DAT) and 6 DAT samples in the control than in the other samples. Among all the known miRNAs, 14 miRNAs showed high abundance > 5000 TPM, including *cpa-miR159a, cpa-miR159b, cpa-miR160d, cpa-miR162a, cpa-miR164a, cpa-miR166a, cpa-miR166d, cpa-miR319, cpa-miR390a, cpa-miR396, cpa-miR477, cpa-miR8137, cpa-miR8141,* and *cpa-miR8148.* Five identified miRNAs from across all the miRNA libraries showed low abundance < 100 TPM, including *cpa-miR160c-3p, cpa-miR164d, cpa-miR5211, cpa-miR8134,* and *cpa-miR8150.* In addition, 169 novel miRNA candidates from 21 different libraries were predicted. It seems that 25.4% of the novel miRNA candidates showed lower abundance < 500 TPM. Among all the identified novel miRNAs, *unconservative_supercontig_106_26263* and *unconservative_supercontig_17_8265* were in greatest abundance (Table S6).

3.3. Identification of Differentially Expressed miRNAs

Compared to 0 DAT, 60 differentially expressed miRNAs (DEMs) ($|\log2(FC)| \geq 1$, p-value ≤ 0.05) were identified (at 1 DAT vs. 0 DAT and 6 DAT vs. 0 DAT) (Figure 2A). There were 46 DEMs identified between 1 and 0 DAT in the control group, of which 40 DEMs showed increased expression and 6 DEMs showed decreased expression (Figure 2A). There were 20 DEMs identified between 6 and 0 DAT in the control group, among which 5 DEMs had increased expression and 15 DEMs had decreased expression. Among the DEMs in both comparison groups, there were six DEMs that were shared across both comparisons during the papaya ripening stage, of which four DEMs had increased expression and two DEMs had decreased expression during fruit ripening. The number of DEMs decreased during fruit ripening (Figure 2A). The enriched GO terms for the predicted target genes are presented in Figure 2B. "Biological process of metabolic process", "single-organism process and cellular process", "the cellular component of cell part and cell", and "the molecular function of catalytic activity and binding" had the most significant differences in gene expression changes during papaya fruit ripening. In the KEGG enrichment analysis, there were 36 top enrichments for metabolic/biological pathways (Figure 2C). Among these enriched genes, "metabolic pathways of plant hormone signal transduction", "fatty acid metabolism", "biosynthesis of amino acids", and "starch and sucrose metabolism" are important for fruit ripening.

Figure 3A shows a total of 60 miRNAs that were differentially expressed between 1-MCP treatments (long-term and short-term) compared to the control (Figure 3A). The amount of DEMs decreased with the storage time and duration of the 1-MCP treatment. Compared to the control, a total of 45 miRNAs were differentially expressed during short-term 1-MCP treatment. There were 35 DEMs specifically identified in the comparison of 1-MCP 2 h vs. CK at 1 DAT, 5 DEMs specifically identified in the comparison of 1-MCP 2 h vs. CK at 6 DAT, and 5 common DEMs shared across both DATs. Contrary to this, a total of 27 DEMs were identified in the comparison of long-term 1-MCP vs. the control. Of these, 11 and 12 DEMs were specifically identified after 1 and 6 DAT, respectively, and 4 DEMs were shared across both DATs. These results indicate that miRNAs are important in regulating fruit ripening.

In the abnormal fruit ripening case caused by 1-MCP treatment, the number of DEMs decreased dramatically compared to short-term 1-MCP treatment, indicating that the missing DEMs may be involved in the softening disorder. In order to fully understand the functions of DEM target genes, a GO enrichment analysis was performed (Figure 3B). The enrichments for "metabolic process and cellular process", "the cellular component of cell part and cell", and "molecular function of catalytic activity and binding" biological processes showed the most significant differences in gene expression during 1-MCP-treated papaya fruit ripening. From the KEGG enrichment terms, there were 38 top enrichments for metabolic/biological pathways (Figure 3C). Among these KEGG enrichment terms, "metabolic pathways of plant hormone signal transduction", "phenylpropanoid biosynthesis", "protein processing in endoplasmic reticulum", "carbon metabolism", and "starch and sucrose metabolism" genes were the most enriched, indicating the important roles of these terms in fruit ripening and 1-MCP-caused ripening disorder. The COG function for different treatments is shown in Figure S1. All DEMs were clustered into 26 functional categories based on four kinds of differential enrichment analyses. Among the 26 functional categories, "general function prediction only", "transcription", and "replication, recombination, and repair" were the most enriched items. The signal transduction mechanism was also enriched, which is an important functional protein for the ripening process of papaya (Figure S1).

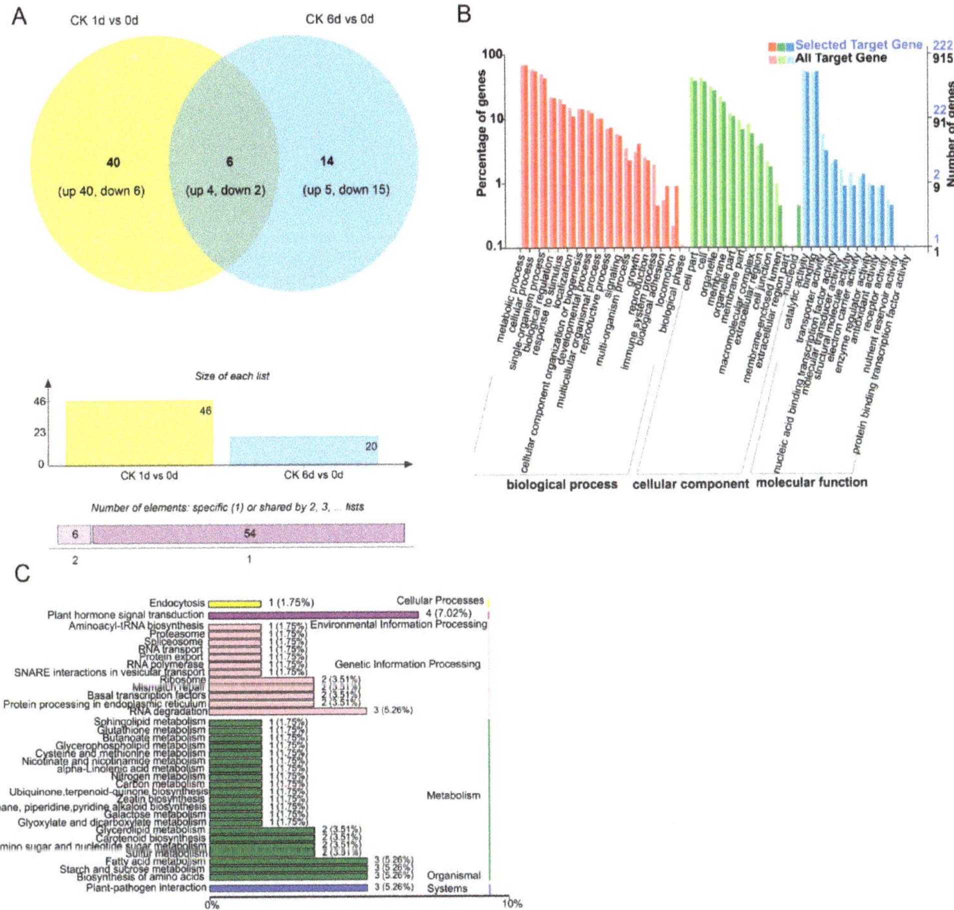

Figure 2. DEM (differentially expressed miRNA) analysis under control conditions during the fruit ripening stage. (**A**) Number of DEMs in papaya fruit at 1 and 6 days after treatment (DAT) compared to 0 DAT under control conditions. The Venn software package (http://bioinformatics.psb.ugent.be/webtools/Venn/) was used for the Venn diagram. (**B**) GO classification of the predicted target genes in the comparison of 1 and 6 DAT with 0 DAT. (**C**) KEGG classification of the predicted targets of target genes in the comparison of 1 and 6 DAT with 0 DAT.

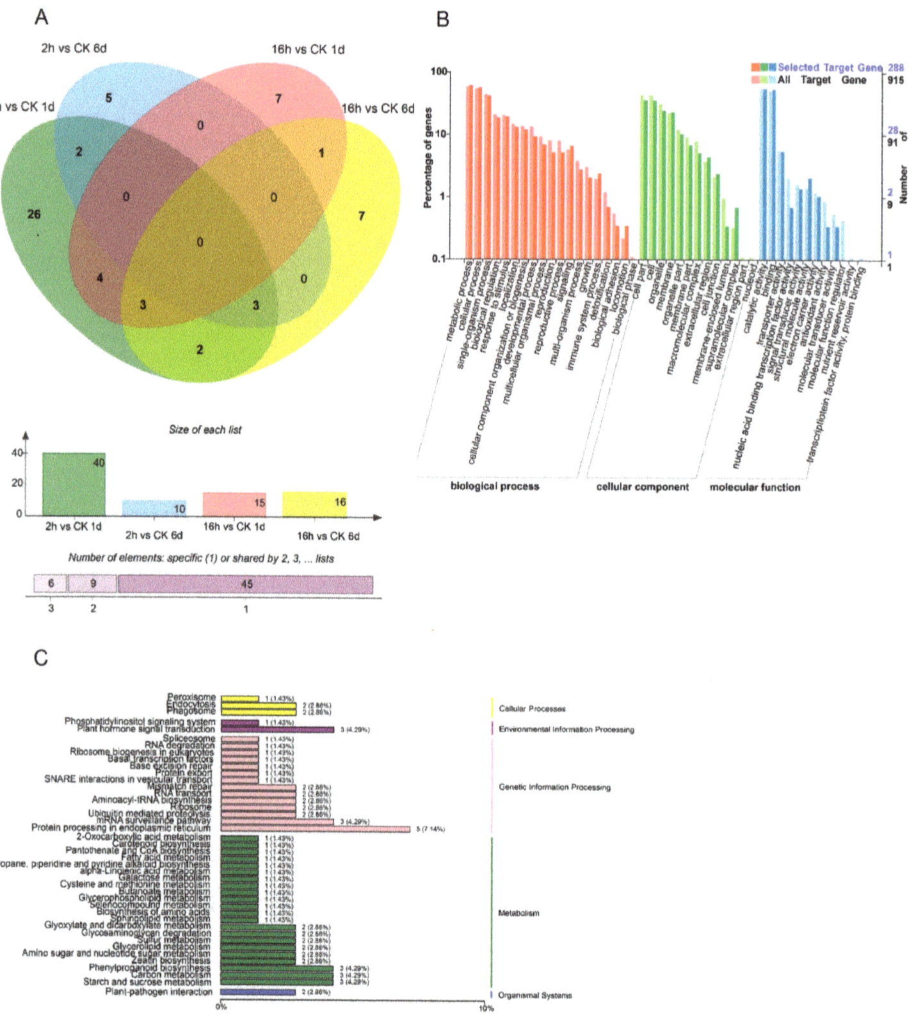

Figure 3. The effect of different 1-MCP treatments on the expression of miRNA. (**A**) The number of DEMs (differential expressed miRNAs) derived from comparison between the 1-MCP treatments (2 h) in 1- and 6-day samples and the control samples at each time point, and the number of DEMs derived from comparison between the 1-MCP treatments (16 h) in 1- and 6-day samples and the control samples at each time point. The Venn software package (http://bioinformatics.psb. ugent.be/webtools/Venn/) was used for the Venn diagram. (**B**) GO classification of the predicted targets of target genes in the 1-MCP (16 h) 1- and 6-day samples compared to the control, and the 1-MCP (2 h) 1- and 6-day samples compared to the control at each time point. (**C**) KEGG classification of the predicted targets of target genes in the 1-MCP (16 h) 1- and 6-day samples compared to the control, and the 1-MCP (2 h) 1- and 6-day samples compared to the control at each time point.

Ethylene and auxin signaling pathways showed an important role in the ripening of papaya treated with different durations of 1-MCP treatment [31]. According to the perfect or near-perfect complementarity of expression between miRNAs and their targets, DEMs were predicted to correspond to uni-genes as potential targets. Through sRNA sequencing analysis, Target Finder software was used for predicting targets. According to sRNA sequencing, miRNAs and their predicted target genes related to ethylene and auxin were found, and eight miRNAs and their predicted target genes with opposite expression levels were screened out. It indicates that miRNAs may play a critical regulatory role in ethylene and auxin signal transduction pathways. The identified miRNAs and their corresponding target genes are shown in Table 1. In the early development stage of papaya ripening (1 DAT vs. 0 DAT in the control), *cpa-miR319* was downregulated, while *cpa-miR396* and *cpa-miR8140* were upregulated. At the late development stage of papaya ripening (6 DAT vs. 0 DAT in the control), the expression of *cpa-miR160d* decreased. At the late development stage of delayed papaya ripening, *cpa-miR390a* (1-MCP 2 h 6 DAT vs. control 6 DAT) and *cpa-miR396* were upregulated; *cpa-miR172a*, on the other hand, was downregulated (1-MCP 2 h 6 DAT vs. 1 DAT). At the early development stage of papaya softening disorder caused by 1-MCP (1-MCP 16 h 1 DAT to control 1 DAT), *cpa-miR172a* was upregulated. Meanwhile, during the ripening stage of papaya with ripening disorder caused by 1-MCP, the expression of *cpa-miR160d, unconservative_supercontig_2_1866* (1-MCP 16 h 6 DAT vs. control 6 DAT), *cpa-miR167c*, and *cpa-miR8140* (1-MCP 16 h 6 DAT vs. 1 DAT) was upregulated. These results indicated that miRNAs may be involved in ethylene and auxin signaling and participate in the ripening of papaya.

3.4. Combined Expression Analysis of miRNAs and Their Target mRNAs

Through differential expression analysis of the miRNAs and mRNAs in the sRNAs and transcriptome sequencing data, key miRNAs and genes were identified. The regulation of miRNAs and their target genes was uncovered. Compared to 0 DAT, 34 DEMs were identified, of which 20 DEMs were found at both 1 and 6 DAT. Six DEMs were shared across both DAT groups (Figure 4A). The enriched target gene GO terms are presented in Figure 4B. The enrichments for "biological process of metabolic process and cellular process" and "the molecular function of binding" biological processes showed the most significant differences in gene expression during papaya fruit ripening. From the KEGG enriched terms, "metabolic pathways of plant hormone signal transduction", "carbon metabolism", and "starch and sucrose metabolism" were important for fruit ripening (Figure 4C).

Combined with transcriptome analysis, a total of 34 miRNAs were expressed in the different 1-MCP treatments (long-term and short-term). The amount of DEMs increased with treatment storage time and duration of the 1-MCP treatment. In total, 22 DEMs were identified in the comparison of short-term 1-MCP vs. the control at 6 and 1 DAT. Among the 22 DEMs, 17 DEMs were identified at 1 DAT, 10 were at identified 6 DAT, and 5 were shared across both dates. The enriched GO terms for miRNA target genes are presented in Figure 5B. The enrichments for "metabolic process and cellular process" and "molecular function of binding" biological processes showed the most significant differences in gene expression during 1-MCP-delayed papaya ripening. From the KEGG enriched terms, "metabolic pathways of plant hormone signal transduction" and "starch and sucrose metabolism" were important for fruit ripening (Figure 5C).

Combined with transcriptome analysis, the miRNA target genes were screened out if they had an opposite expression pattern to their respective miRNAs. In total, 12 miR-NAs were found to be involved in the signal transduction process of ethylene and auxin. The miRNAs and their corresponding target genes are shown in Table 2. At the early development stage of papaya ripening (1 DAT vs. 0 DAT in the control group), the expression of *cpa-miR319* was downregulated while *cpa-miR396, cpa-miR8140, unconservative_supercontig_120_28048*, and *unconservative_supercontig_46_16464* were upregulated. At the late development stage of papaya ripening (6 DAT vs. 0 DAT in the control), the expression of *cpa-miR160d* and *unconservative_supercontig_2_1866* was downregulated while that of *unconservative_supercontig_46_16464* was upregulated. At the late ripening stage of papaya fruit under suitable 1-MCP treatment (1-MCP 2 h 6 DAT vs. control 6 DAT), the expression of *cpa-miR390a* was upregulated. During the ripening stage of papaya fruit under suitable 1-MCP treatment, the expression of *cpa-miR172a* (1-MCP 2 h 6 DAT vs. 1 DAT) was upregulated, while the expression of *cpa-miR396* and *unconservative_supercontig_9_5033* was downregulated (1-MCP 2 h 6 DAT to 1 DAT). At the early development stage in papaya with the ripening disorder caused by 1-MCP (1-MCP 16 h 1 DAT vs. control 1 DAT), the expression of *cpa-miR172a* and *unconservative_supercontig_52_17983* was upregulated, and that of *unconservative_supercontig_120_28048* was downregulated. At the late development stage in papaya with ripening disorder, *cpa-miR160d* and *unconservative_supercontig_2_1866* were upregulated (1-MCP 16 h 6 DAT to control 6 DAT). During the papaya ripening stage for individuals with the ripening disorder, the expression of *cpa-miR167c* and *cpa-miR8140* was upregulated (1-MCP 16 h 6 DAT to 1 DAT).

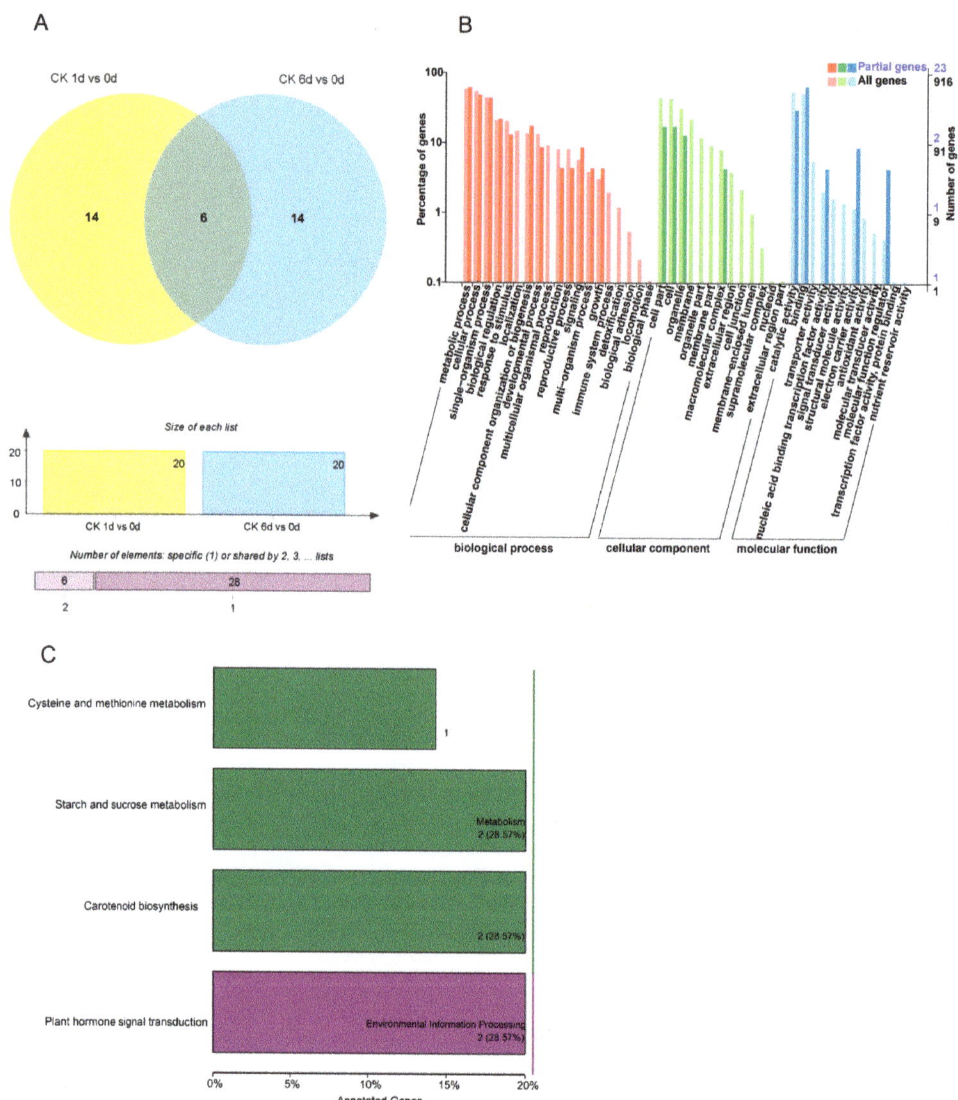

Figure 4. Changes in DEMs (differentially expressed miRNAs) during papaya fruit ripening from both miRNA sequencing and transcriptomic analysis. (**A**) The number of DEMs in papaya fruit at 1 and 6 days after treatment (DAT) compared to 0 DAT under control conditions. (**B**) GO classification of target gene comparison at 1 and 6 DAT compared to 0 DAT. (**C**) KEGG classification of target gene comparison at 1 and 6 DAT compared to 0 DAT.

Table 1. Expression profiles of DEMs and the predicted target genes under different conditions.

	miRNA	miR_Regulate	Predicted Target Gene ID	Predicted Target Gene Name
CK 1 d vs. 0 d	cpa-miR319	Down	evm.TU.supercontig_7.3	*Carica papaya* auxin response factor 3
	cpa-miR396	Up	evm.TU.supercontig_48.26	*Carica papaya* 1-aminocyclopropane-1-carboxylate synthase 1-like
	cpa-miR396	Up	evm.TU.supercontig_481.1	*Carica papaya* ethylene-responsive transcription factor RAP2-12-like
	cpa-miR8140	Up	evm.TU.supercontig_1322.1	*Carica papaya* 1-aminocyclopropane-1-carboxylate synthase 1-like
CK 6 d vs. 0 d	cpa-miR160d	Down	evm.TU.supercontig_49.122	*Carica papaya* auxin response factor 17
	cpa-miR160d	Down	evm.TU.supercontig_53.88	*Carica papaya* auxin response factor 16-like
	cpa-miR160d	Down	evm.TU.supercontig_65.4	*Carica papaya* auxin response factor 10
2 h 6 d vs. CK 6 d	cpa-miR390a	Up	evm.TU.supercontig_261.2	*Carica papaya* auxin response factor 19-like
16 h 1 d vs. CK 1 d	cpa-miR172a	Up	C.papaya_newGene_850	*Carica papaya* auxin transport protein BIG
	cpa-miR172a	Up	evm.TU.supercontig_1.271	*Carica papaya* ethylene-responsive transcription factor RAP2-7-like
	cpa-miR172a	Up	evm.TU.supercontig_114.55	*Carica papaya* ethylene-responsive transcription factor RAP2-7-like
16 h 6 d vs. CK 6 d	cpa-miR160d	Up	evm.TU.supercontig_49.122	*Carica papaya* auxin response factor 17
	cpa-miR160d	Up	evm.TU.supercontig_53.88	*Carica papaya* auxin response factor 16-like
	cpa-miR160d	Up	evm.TU.supercontig_65.4	*Carica papaya* auxin response factor 10
	unconservative_supercontig_2_1866	Up	evm.TU.supercontig_7.3	*Carica papaya* auxin response factor 3
2h 6 d vs. 1 d	cpa-miR172a	Up	C.papaya_newGene_850	*Carica papaya* auxin transport protein BIG
	cpa-miR172a	Up	evm.TU.supercontig_1.271	*Carica papaya* ethylene-responsive transcription factor RAP2-7-like
	cpa-miR172a	Up	evm.TU.supercontig_114.55	*Carica papaya* ethylene-responsive transcription factor RAP2-7-like
	cpa-miR396	Down	evm.TU.supercontig_48.26	*Carica papaya* 1-aminocyclopropane-1-carboxylate synthase 1-like
	cpa-miR396	Down	evm.TU.supercontig_481.1	*Carica papaya* ethylene-responsive transcription factor RAP2-12-like
16 h 6 d vs. 1 d	cpa-miR167c	Up	evm.TU.supercontig_17.52	*Carica papaya* auxin response factor 6
	cpa-miR8140	Up	evm.TU.supercontig_1322.1	*Carica papaya* 1-aminocyclopropane-1-carboxylate synthase 1-like

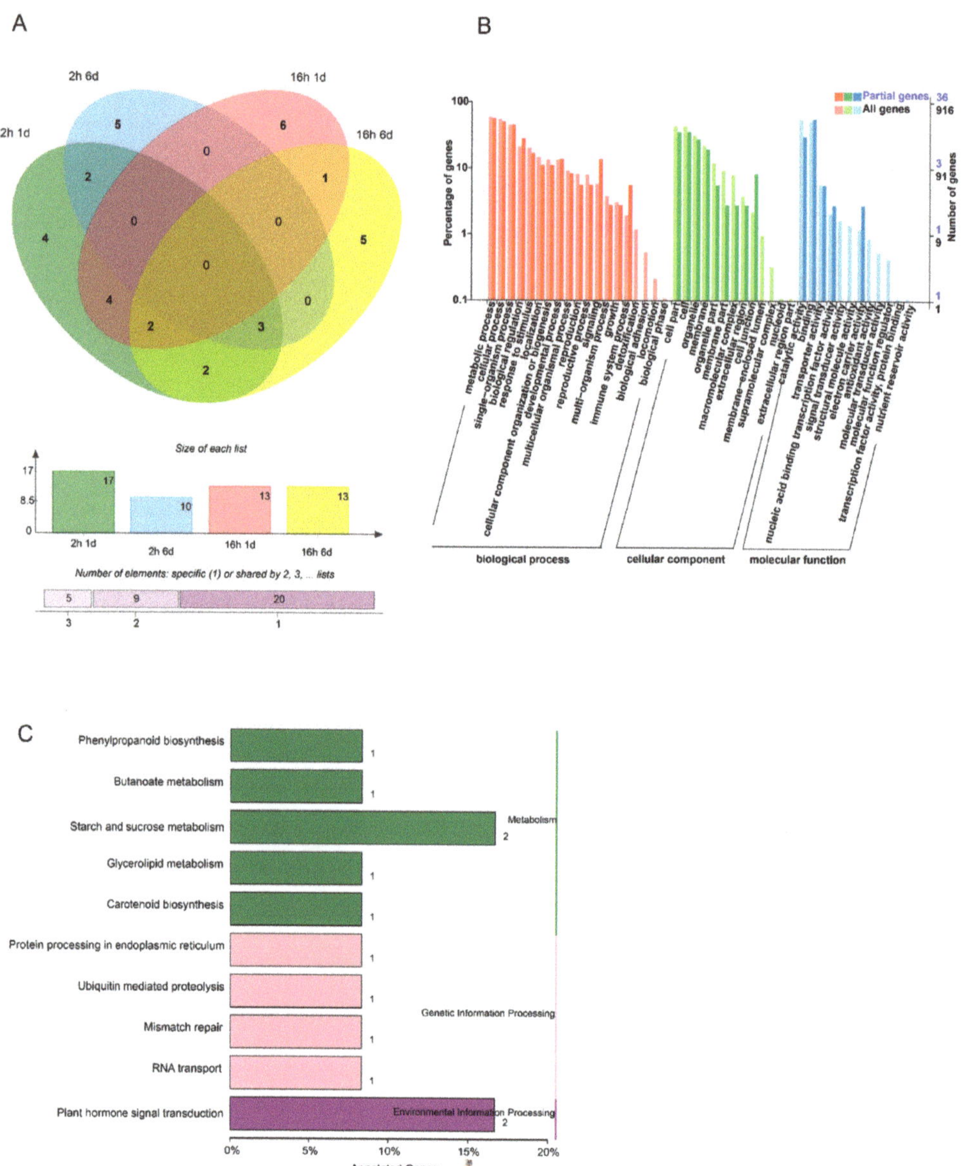

Figure 5. The effect of 1-MCP treatments on both miRNA sequence and transcriptomic analysis. (**A**) The number of DEMs (differentially expressed miRNAs) derived from comparison between the 1-MCP (2 h) treatment and the control on 1- and 6-day samples, and the number of DEMs derived from comparison between the 1-MCP (16 h) treatment and the control on 1- and 6-day samples. (**B**) GO classification of target genes between the 1-MCP (16 h) 1- and 6-day samples and the 1-MCP (2 h) samples at each time point. (**C**) KEGG classification of target genes between the 1-MCP (16 h) 1- and 6-day samples and the 1-MCP (2 h) samples at each time point.

Table 2. Expression profiles of DEMs and their target genes under small RNAs and transcriptome sequencing.

	miRNA	miR_Regulate	Target Gene ID	Target Gene Name	Target Gene Regulation
CK 1 d vs. 0 d	cpa-miR396	Up	evm.TU.supercontig_48.26	*Carica papaya* 1-aminocyclopropane-1-carboxylate synthase 1-like	Down
	cpa-miR396	Up	evm.TU.supercontig_481.1	*Carica papaya* ethylene-responsive transcription factor RAP2-12-like	Down
	cpa-miR8140	Up	evm.TU.supercontig_1322.1	*Carica papaya* 1-aminocyclopropane-1-carboxylate synthase 1-like	Down
	unconservative_supercontig_120_28048	Up	evm.TU.contig_32826	*Carica papaya* indole-3-acetic acid-amido synthetase GH3.6	Down
	unconservative_supercontig_46_16464	Up	evm.TU.supercontig_2.209	*Carica papaya* 1-aminocyclopropane-1-carboxylate oxidase homolog 4-like	Down
	cpa-miR319	Down	evm.TU.supercontig_7.3	*Carica papaya* auxin response factor 3	Up
CK 6 d vs. 0 d	unconservative_supercontig_46_16464	Up	evm.TU.supercontig_2.209	*Carica papaya* 1-aminocyclopropane-1-carboxylate oxidase homolog 4-like	Down
	cpa-miR160d	Down	evm.TU.supercontig_49.122	*Carica papaya* auxin response factor 17	Up
	cpa-miR160d	Down	evm.TU.supercontig_53.88	*Carica papaya* auxin response factor 16-like	Up
	cpa-miR160d	Down	evm.TU.supercontig_65.4	*Carica papaya* auxin response factor 10	Up
	unconservative_supercontig_2_1866	Down	evm.TU.supercontig_7.3	*Carica papaya* auxin response factor 3	Up
2h 6 d vs. CK 6 d	cpa-miR390a	Up	evm.TU.supercontig_261.2	*Carica papaya* auxin response factor 19-like	Down
16 h 1 d vs. CK 1 d	cpa-miR172a	Up	C.papaya_newGene_850	*Carica papaya* auxin transport protein BIG	Down
	cpa-miR172a	Up	evm.TU.supercontig_1.271	*Carica papaya* ethylene-responsive transcription factor RAP2-7-like	Down
	cpa-miR172a	Up	evm.TU.supercontig_114.55	*Carica papaya* ethylene-responsive transcription factor RAP2-7-like	Down
	cpa-miR172a	Up	evm.TU.supercontig_139.43	*Carica papaya* ethylene-responsive transcription factor RAP2-7	Down
	unconservative_supercontig_120_28048	Down	evm.TU.contig_32826	*Carica papaya* indole-3-acetic acid-amido synthetase GH3.6	Up
	unconservative_supercontig_52_17983	Up	evm.TU.supercontig_83.80	*Carica papaya* ethylene-responsive transcription factor 4	Down

Table 2. *Cont.*

16 h 6 d vs. CK 6 d	cpa-miR160d	Up	evm.TU.supercontig_49.122	*Carica papaya* auxin response factor 17	Down
	cpa-miR160d	Up	evm.TU.supercontig_53.88	*Carica papaya* auxin response factor 16-like	Down
	cpa-miR160d	Up	evm.TU.supercontig_65.4	*Carica papaya* auxin response factor 10	Down
	unconservative_supercontig_2_1866	Up	evm.TU.supercontig_7.3	*Carica papaya* auxin response factor 3	Down
2 h 6 d vs. 1 d	cpa-miR172a	Up	C.papaya_newGene_850	*Carica papaya* auxin transport protein BIG	Down
	cpa-miR172a	Up	evm.TU.supercontig_1.271	*Carica papaya* ethylene-responsive transcription factor RAP2-7-like	Down
	cpa-miR172a	Up	evm.TU.supercontig_114.55	*Carica papaya* ethylene-responsive transcription factor RAP2-7-like	Down
	cpa-miR172a	Up	evm.TU.supercontig_139.43	*Carica papaya* ethylene-responsive transcription factor RAP2-7	Down
	cpa-miR396	Down	evm.TU.supercontig_48.26	*Carica papaya* 1-aminocyclopropane-1-carboxylate synthase 1-like	Up
	cpa-miR396	Down	evm.TU.supercontig_481.1	*Carica papaya* ethylene-responsive transcription factor RAP2-12-like	Up
	unconservative_supercontig_9_5033	Down	evm.TU.supercontig_9.126	*Carica papaya* auxin-binding protein T85	Up
16 h 6 d vs. 1 d	cpa-miR167c	Up	evm.TU.supercontig_17.52	*Carica papaya* auxin response factor 6	Down
	cpa-miR8140	Up	evm.TU.supercontig_1322.1	*Carica papaya* 1-aminocyclopropane-1-carboxylate synthase 1-like	Down

3.5. Target Gene Identification of Papaya miRNAs by Degradome Analysis

Through degradome sequencing, the splice sites of miRNAs in the mRNA were found. The 1741 target genes corresponding to their miRNAs are summarized in Table S7. According to KEGG classification, the target genes of these key DEMs related to plant hormone signal transduction are summarized and shown in Figure 6A,B. The plant hormone signal pathways include ethylene, jasmonic acid, auxin, gibberella, MAPK, and abscisic acid. Among all these pathways, the miRNAs related specifically to the ethylene and auxin signaling pathways were the most enriched. The corresponding key target genes were *CpCTR1*, *CpARFs*, *CpTIR1*, and *CpSAUR67*. The expression of some miRNAs increased with fruit ripening, such as *unconservative_supercontig_106_26263*, *cpa-miR156a*, and *cpa-miR160d* and *cpa-miR160a*. Some miRNAs were closely related to papaya softening disorder, such as *unconservative_supercontig_20_9672*, *unconservative_supercontig_13_6816*, *cpa-miR8148*, *unconservative_supercontig_152_30952*, and *unconservative_supercontig_119_27967*. Four miRNA targets to *CpCTR1* were identified: *unconservative_supercontig_75_21810*, *unconservative_supercontig_2_1866*, *unconservative_supercontig_19_9148*, and *unconservative_supercontig_3_2646*. *CpSAUR67* and *CpTIR1* were involved in auxin biosynthesis and signal transduction and were putative targets for *unconservative_supercontig_20_9672* and *cpa-miR393*, respectively. *CpARFs* were putative targets for *cpa-miR160d*, *cpa-miR319*, *unconservative_supercontig_184_32956*, *cpa-miR156a*, and *cpa-miR160a*. The expression of these target genes was neither increased with fruit ripening and repressed by 1-MCP treatment nor decreased with fruit ripening, but it was induced by 1-MCP treatments (Figure 6B), indicating that these target genes may play an important role in the fruit ripening process.

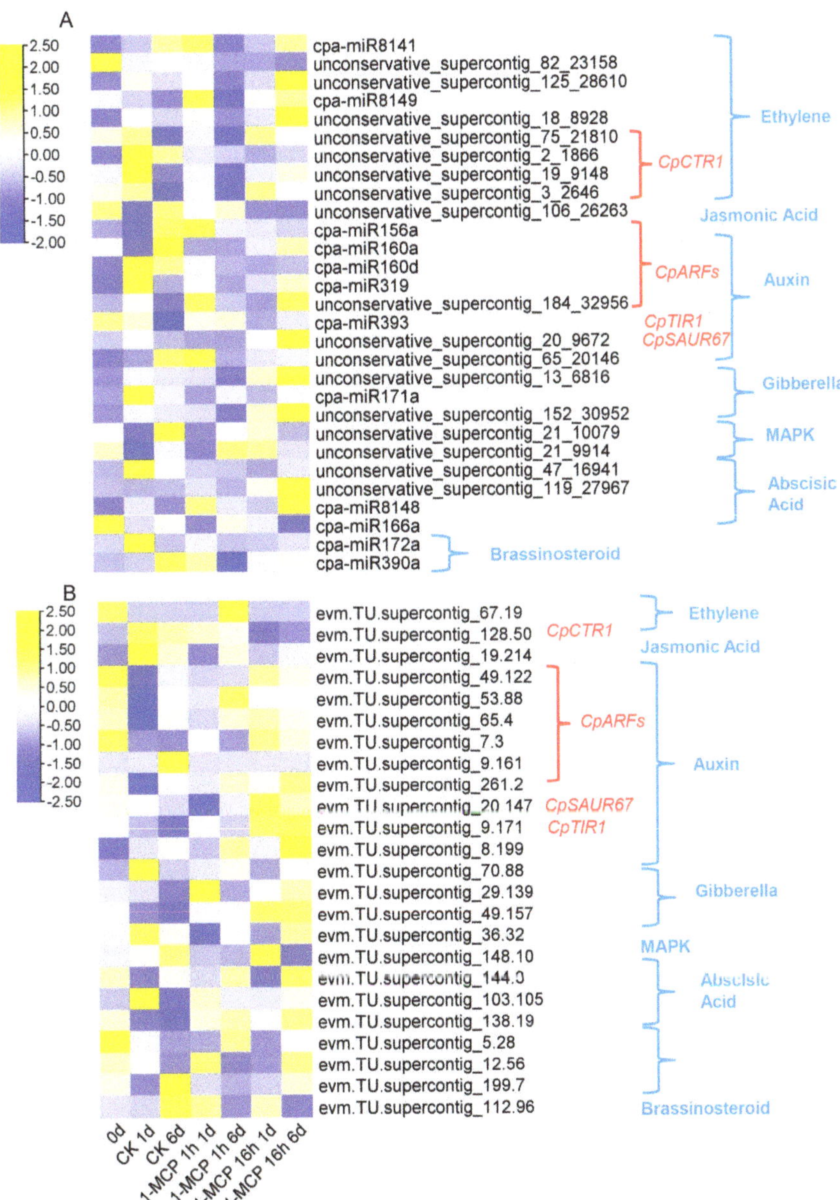

Figure 6. Expression profiles of DEMs (differentially expressed miRNAs) and their candidate target genes in hormone signal pathways. (**A**) Heat map display of DEM expression profiles involved in plant hormone signaling metabolic pathways under 1-MCP treatment combined with the degradation analysis. (**B**) Heat map display of the candidate target genes of DEMs involved in plant hormone signaling metabolic pathways under 1-MCP treatment combined with the degradation analysis.

3.6. Target Gene Identification of Papaya miRNAs by Degradome Analysis

According to the analysis of small RNAs, degradome, and transcriptome sequencing, the miRNAs related to ethylene and auxin signaling were selected and visualized. The *cpa-miR160a* and *cpa-miR160d* miRNAs co-targeted *CpARF10*, *CpARF16-like*, and *CpARF17*, thereby affecting auxin signaling and fruit ripening. The *cpa-miR172a* miRNA targeted *CpERF RAP2-7* and *Carica papaya auxin transport protein BIG*, and *unconservative_supercontig_2_1866* targeted *CpARF3* and *CpCTR1*, thereby affecting the signaling pathways of ethylene and auxin simultaneously (Figure 7).

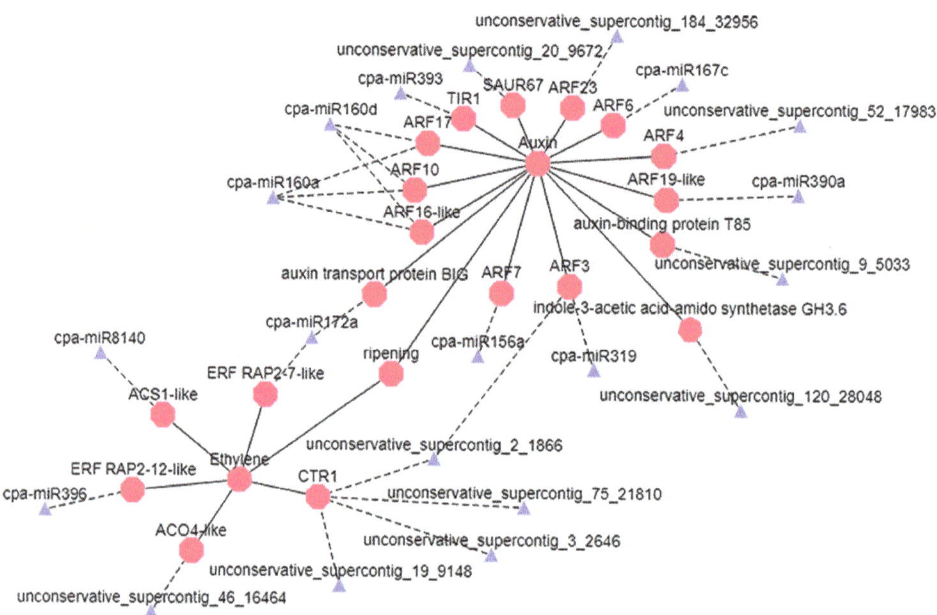

Figure 7. Regulation network diagram of miRNAs and their candidate target genes. Blue triangles represent the miRNAs, pink polygons represent the corresponding target genes, the dotted line shows negative regulation, and the straight line shows positive regulation.

3.7. Target Gene Identification of Papaya miRNAs by Degradome Analysis

Through the analysis of sRNAs and mRNA sequencing, the miRNAs and their target genes that are related to ethylene and auxin were selected and verified by RT-qPCR. There were four miRNAs related to ethylene signaling, namely *cpa-miR172a*, *cpa-miR396*, *cpa-miR8140*, and *unconservative_supercontig_46_16464*. Among these miRNAs, the expression of *miR172a* in the control group increased sharply at 1 DAT and then decreased. 1-MCP treatments severely repressed the expression of *miR172a* (Figure 8A). *CpERF RAP2-7-like* was predicted to be the target gene of *miR172a*. The expression of *CpERF RAP2-7-like* decreased at 1 DAT and then increased at 6 DAT, with an opposite expression trend compared to *miR172a* (Figure 8B). The expression of *cpa-miR396*, *cpa-miR8140*, and *unconservative_supercontig_46_16464* decreased with fruit ripening, while the expression of their corresponding target genes, namely *CpERF RAP2-12-like*, *CpACS1-like*, and *CpACO4-like*, increased first and then decreased, respectively (Figure 8C–H). Among the four ethylene-related genes, the expression level of *CpACS1* significantly increased at 1 DAT in the control sample, signaling that these genes may play an important role in the early development stage of fruit ripening. The expression levels of *CpACO4*, *CpERF4*, and *CpERF RAP2-7-like* slightly decreased and were negatively correlated with fruit ripening. The expression of

CpERF RAP2-12-like increased with fruit ripening and was positively correlated with fruit ripening (Figure 8). The RT-qPCR results were consistent with the RNA-seq data, and miRNA and the corresponding target genes showed opposite expression patterns.

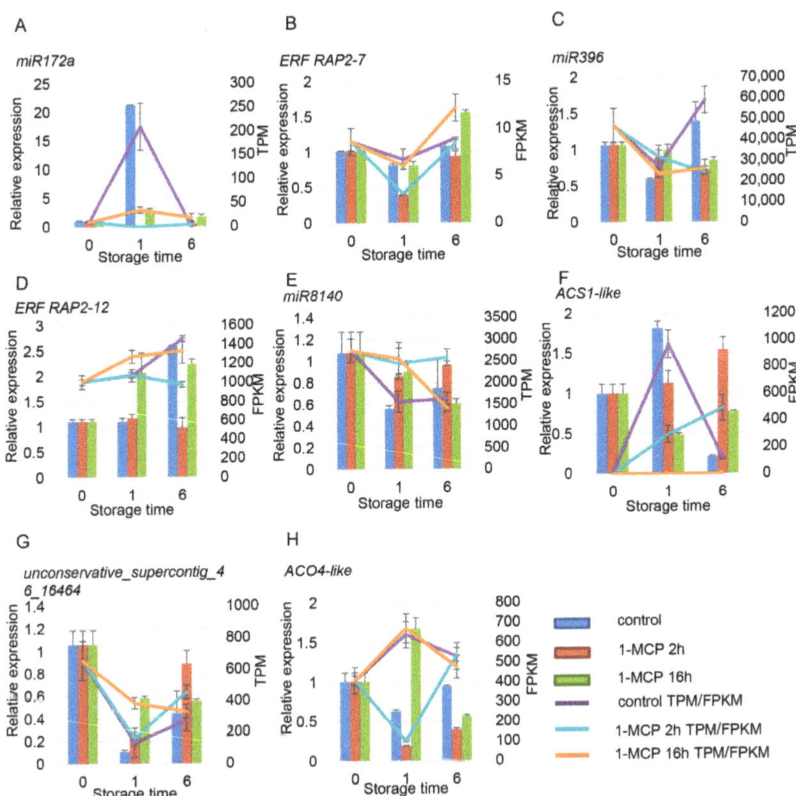

Figure 8. Expression pattern validation of selected DEMs (differentially expressed miRNAs) and their target genes related to the ethylene signaling pathway under different 1-MCP treatments. The histograms were plotted using data obtained from RT-qPCR, and the corresponding line chart was plotted using TPM/FPKM values from the RNA-seq analysis. (**A–H**) are miRNA and the their target gene pairs. (**A,C,E,G**) are the miRNA, and (**B,D,F,G**) are the corresponding target gene, respectively. Different colors indicate different samples. The expression of samples at 0 DAT was set to 1. *ACT* and *TBP1* were used as references, which were validated by Zhu et al. (2012).

3.8. Target Gene Identification of Papaya miRNAs by Degradome Analysis

There were six miRNAs and seven target genes identified related to auxin signaling. Among the miRNAs and target genes, miR160a and miR160d worked together to regulate the expression of *ARF10/16-like/17*. Similar expression profiles in *miR160d, miR167c, miR319, miR390a*, and *unconservative_supercontig_2_1866* were observed, which first increased and then decreased (Figure 9A,E,G,K). On the contrary, expression of the target gene ARF first decreased and then increased (Figure 9B–D,F,H,L). The expression of *unconservative_supercontig_9_5033* first decreased and then increased; the expression of its target gene, *auxin-binding protein T85*, increased first and then decreased (Figure 9I,J). The expression of all six of these miRNAs was positively correlated with fruit ripening, but their targets showed a negative correlation with fruit ripening. 1-MCP treatment repressed the expression of these miRNAs but induced the expression of their target genes (Figure 9).

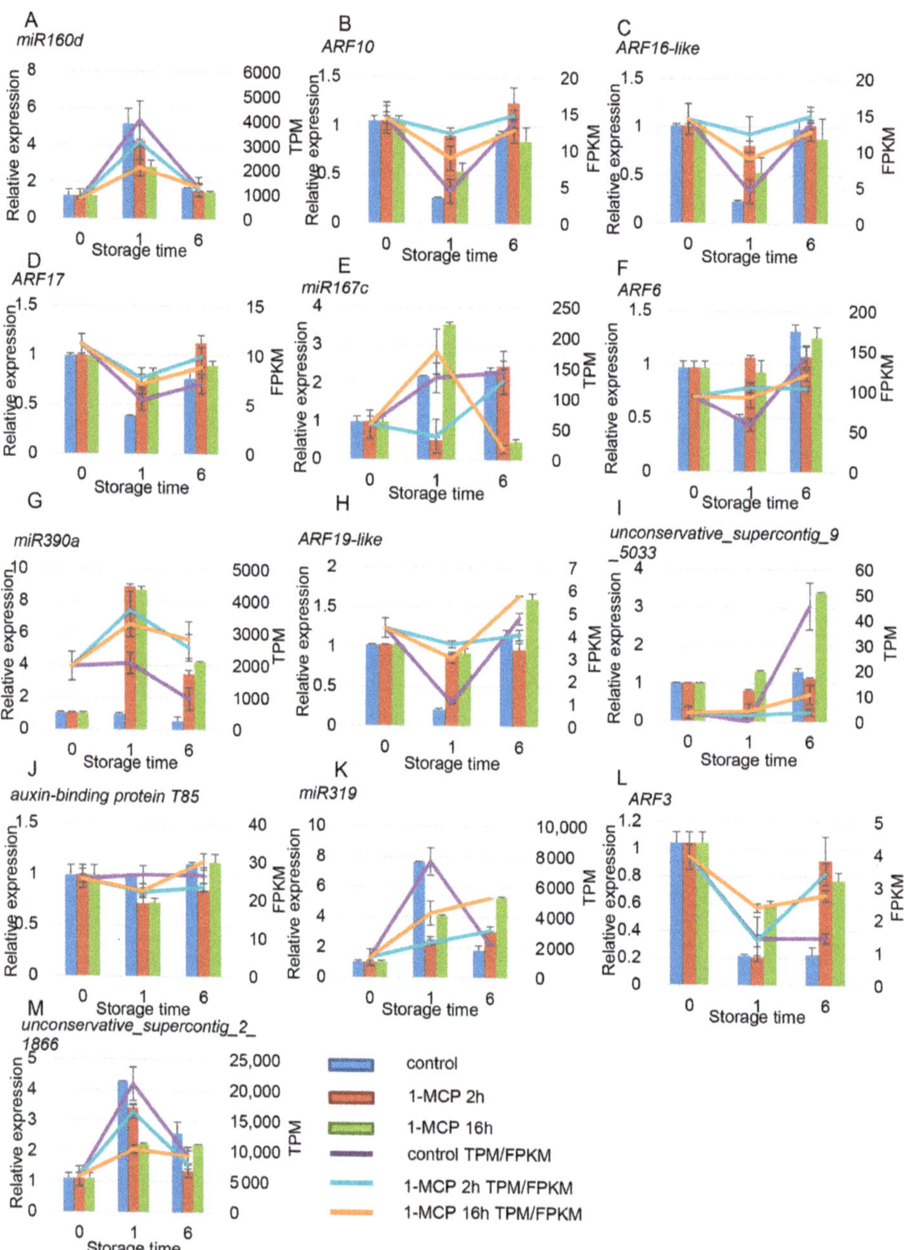

Figure 9. Expression pattern validation of selected DEMs (differentially expressed miRNAs) and their target genes related to the auxin signaling pathway under different 1-MCP treatments. The histograms were plotted using data obtained from RT-qPCR, and the corresponding line chart was plotted using TPM/FPKM values from the RNA-seq analysis. (**B–D**) are the target genes of (**A**); (**F**) is the target genes of (**E**); (**H**) is the target genes of (**G**); (**J**) is the target genes of (**I**); (**L**) is the target genes of (**K,M**). Different colors indicate different samples. The expression of samples at 0 DAT was set to 1. *ACT* and *TBP1* were used as references, which were validated by Zhu et al. (2012).

3.9. Target Gene Identification of Papaya miRNAs by Degradome Analysis

According to the previous analysis, *CpARF19-like* and *CpERF RAP2-12-like* were potential targets of cpa-miR390a and cpa-miR396, respectively. To further test the relationship between these miRNAs and their target genes, transient co-expression assays in *Nicotiana benthamiana* leaves were conducted. As shown in Figure 10, the green fluorescent protein (GFP) fluorescence from tobacco leaf areas injected with 35s::Pre-cpa-miR390a + 35s::CpARF19-like-GFP and 35s::Pre-cpa-miR396 + 35s::CpERF RAP12-2-like-GFP was starkly lower than that of the negative controls (Figure 10). These results provide evidence that the *CpARF19-like* and *CpERF RAP2-12-like* genes may be directly regulated by cpa-miR390 and cpa-miR396 miRNA, respectively.

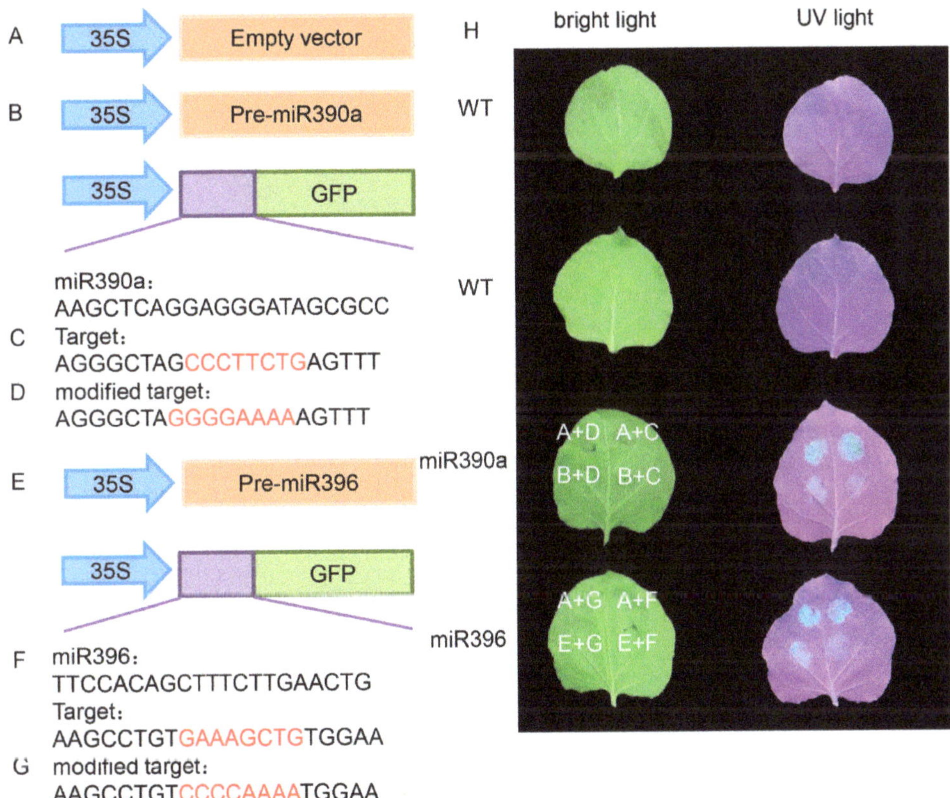

Figure 10. Confirmation of miRNAs and their targets in tobacco. (**A–G**) Diagrams of the plasmids used in these experiments. (**A**) Empty vector; (**B**) 35s::pre-miR390a vector that overexpresses Cpa-miR390a; (**C**) 35s::CpARF 19-like-GFP vector that overexpresses CpARF19-like; (**D**) 35s::MCpARF 19-like-GFP vector that overexpresses GFP carrying a modified target; (**E**) 35s::pre-miR396 vector that overexpresses Cpa-miR396; (**F**) 35s::CpERF RAP2-12-like-GFP vector that overexpresses CpERF RAP2-12-like; (**G**) 35s::MCpARF RAP2-12-like-GFP vector that overexpresses GFP carrying a modified target; (**H**) GFP picture of tobacco.

4. Discussion

4.1. Papaya miRNAs with Conserved and New Gene Targets

As an inhibitor of ethylene receptors, 1-MCP can effectively delay fruit ripening. This has wide applicability to post-harvest fruit preservation. Improper 1-MCP treatment can negatively affect the softening ability of fruit during ripening, resulting in papaya ripening disorder [37–39]. 1-MCP function can reduce the production of endogenous ethylene. In banana, 12 novel and 128 known miRNAs were identified. Among these miRNAs, 22 were differentially expressed after ethylene regulation treatment and were involved in the ethylene response [30]. In Wang's (2020) study between wild-type and *LeERF1* transgenic tomato fruits, 9 miRNAs and 12 nat-siRNAs were found to be differentially expressed [40]. Degradome sequencing analysis further validated the nine miRNA targets, and six new targets were also identified. In our present study, 46 known miRNA families and 169 papaya-specific miRNAs were identified in 1-MCP-treated papaya samples using deep sequencing and computational analyses (Figure 1, Table S5). This difference in these two studies may be due to the type of fruit and the difference in post-harvest treatments. In papaya, a total of 1741 target genes of 178 known miRNAs were identified through degradome analysis (Table S7). Across the identified miRNAs, there were 40 known miRNAs, 138 novel miRNAs, and 29 miRNAs related to plant hormone signaling.

4.2. miRNAs Participate in Hormone Pathways during Papaya Fruit Ripening

Using miRNA sequencing analysis, 34 miRNAs related to papaya ripening and 60 miRNAs related to 1-MCP treatment were found (Figures 2 and 3). After target prediction (Table 1) and combined analysis of miRNAs and miRNA sequences, 12 miRNAs were found to directly act on ethylene- and auxin-related pathways. Of these 12 miRNAs, eight were known miRNAs and one was a novel miRNA. miRNAs are negative regulators of expression of their target genes. The miRNA mdm-miR160 targets the auxin response factors *MdARF16* and *MdARF17* and participates in the formation of adventitious root in apple rootstocks [27]. During papaya's ripening stage (6 DAT vs. 0 DAT in the control), the expression of *cpa-miR160d* decreased, while in papaya fruit with abnormal ripening (1-MCP 16 h 6 DAT vs. control 6 DAT), the expression of *cpa-miR160d* increased. These results indicate that *cpa-miR160d* is involved in the ripening process of papaya. The miRNA miR396 regulates plant growth and development by inhibiting the expression of growth-regulating factor (GRF) [41], participates in the development of wheat plants [42], and responds to environmental stressors such as cold injury, high temperature, and drought [43–45]. In our study, during normal fruit ripening (1 DAT vs. 0 DAT in the control), the expression of *cpa-miR396* was found to be upregulated, while at the early development stage of 1-MCP-delayed ripening (1-MCP 2 h 1 DAT vs. control 1 DAT) in papaya, the expression of *cpa-miR396* was downregulated. In Luan's (2018) study, by overexpressing *miR172a/b* in tomato plants, the expression of *AP2/ERF* transcription factor was inhibited, and the chlorophyll content and photosynthetic rate increased, as well as the development of higher resistance to phytophthora infection [46]. In the present work, the expression of *cpa-miR172a* showed no significant changes during normal ripening (1 DAT vs. 0 DAT in the control), while during the delayed papaya ripening process (1-MCP 2 h 6 DAT vs. 1 DAT) and the early development stage with abnormal ripening (1-MCP 16 h 1 DAT to control 1 DAT), the expression of *cpa-miR172a* was found to be upregulated. These results indicate that the identified miRNAs may play key roles in papaya fruit ripening.

4.3. Degradation Analysis Showed That miRNAs Are Involved in Regulation of Ethylene and Auxin Signaling Pathways

Through miRNA-targeted degradome analysis, miRNAs related to ethylene and auxin signaling were identified, as shown in Figure 6A,B. Their corresponding key target genes are included in Figure 6B: *CpCTR1*, *CpARFs*, *CpTIR1*, and *CpSAUR67*. The miRNAs and their corresponding target genes identified by degradome analysis are shown in Table S8. These miRNAs were found to regulate hormone signal transduction pathways and hormone home-ostasis related to developmental processes. CTR1 is located downstream of the ethylene receptor and mediates the signal from the ethylene receptor by negatively regulating the ethylene response. Both *miR1917* and *miR171b* target *CTR1* and are important regulators for ethylene signal transduction in the ripening of tomato fruit [47,48]. In the present study, *CTR1* was targeted by *unconservative_supercontig_75_21810*, *unconservative_supercontig_2_1866*, *unconservative_supercontig_19_9148*, and *unconservative_supercontig_3_2646*. The expression patterns of *unconservative_supercontig_75_21810*, *unconservative_supercontig_19_9148*, and *unconservative_supercontig_3_2646* decreased during fruit ripening. The expression of *unconservative_supercontig_2_1866* decreased in 1-MCP-delayed fruit ripening.

At the early stages of auxin signal transduction, it was found that several gene families, including Aux/IAA (auxin/indole-3-acetic acid), SAUR (small auxin up RNA), and GH3 (Gretchen Hagen3), could respond to auxin treatments. The Aux/IAA protein acts as an important component of the auxin signaling pathway by participating in the regulation of expression for a large number of genes downstream of the auxin signaling pathway by releasing auxin response factor (ARF). Most of the Aux/IAA proteins contain four conserved domains, namely the I, II, III, and IV domains. Domain II is a key component that leads to instability of the Aux/IAA protein, which is later degraded by the ubiquitin-proteasome protein (TIR1) pathway [49,50]. Therefore, TIR1, SAUR, and ARF are key factors in the auxin signaling pathway. In *Arabidopsis thaliana*, *miR393* participates in auxin signal transduction and plant development by regulating *TIR1* [50]. The *miR165/166* miRNA determines the fate of *A. thaliana* root cells, is involved in plant hormone cross-talk, and regulates root growth via negative regulation of its target genes, such as the auxin response factors *ARF10*, *ARF16*, and *ARF17* [51]. In the present study, *ARFs* were targeted by *miR156a*, *miR160a*, *miR160d*, *miR319*, and *unconservative_supercontig_182_32956*. The expression of these five miRNAs decreased during fruit ripening. *TIR1* and *SAUR67* were targeted by *miR393* and *unconservative_supercontig_65_20146*, respectively. The expression of *miR393* decreased while that of *unconservative_supercontig_65_20146* increased during papaya fruit ripening.

4.4. Network Regulation Diagram of miRNAs and Target Gene Verification

According to the regulatory network diagram, RT-qPCR verification of key miRNAs and target genes was performed, showing that the expression of *cpa-miR172a*, *cpa-miR319*, *cpa-miR390a*, and *cpa-miR396* greatly changed. Previous studies found that miR172a-mediated upregulation of *ERF RAP2-7* during water submergence stress in maize roots restricted plant growth during flood-stress conditions [52]. In the present study, *cpa-miR172a* had an effect on ethylene and auxin signaling pathways. Here, *cpa-miR396* acted on the ethylene signaling pathway, and *cpa-miR319* and *cpa-miR390a* acted on the auxin signaling pathway. Similar results also showed that miR390 targeted ARFs and repressed their expression [53]. It has previously been reported that miR390 and ARFs form an auxin-responsive regulatory network to control lateral root growth in *A. thaliana*. The expression of miR390 is confined to the mesenchymal cells of the xylem prior to lateral root initiation. It was found that miR390 stimulates the production of tasiARFs, which repress the expression of their targets ARF3 and ARF4. There was also positive and negative feedback between ARF2/ARF3/ARF4 and miR390 to regulate lateral root growth [54]. Two highly expressed miRNAs were selected from the ethylene and auxin pathways, and their target genes were verified. We found that the expression signal of GFP protein with target gene binding in tobacco was weakened. These results indicate that *CpARF19-like* and *CpERF RAP2-12-like* are potential targets of cpa-miR390a and cpa-miR396, respectively, and they all may be important in the process of papaya fruit ripening.

5. Conclusions

Suitable 1-MCP treatment effectively delays the ripening of papaya fruit, and long-term 1-MCP treatment causes papaya ripening disorder. In this study, after different 1-MCP treatments on papaya fruits at different development stages, miRNA, transcriptome, and degradome analyses were performed. A total of 213 miRNAs and 1741 target genes of these miRNAs were identified. Among these, 11 different miRNAs related to ethylene and auxin and 12 corresponding target genes were found. The analysis and verification of *cpa-miR390a* and *cpa-miR396*, targeting *CpARF19-like* and *CpERF RAP2-12-like*, respectively, highlighted that these miRNAs and their target genes are likely partners in the ethylene and auxin signaling pathways. Our results indicate that these miRNAs may play an important role in regulating fruit ripening by targeting ethylene and auxin signaling pathways. Unsuitable 1-MCP treatment may disruptively repress miRNA function and cause fruit ripening disorder.

Supplementary Materials: The following are available online at https://www.mdpi.com/article/10.3390/foods10071643/s1, Figure S1: Histogram of COG classification; Table S1: RT-qPCR primers of miRNAs; Table S2: RT-qPCR primers of target genes; Table S3: Primers for plant expression vectors; Table S4: Sequencing data statistics table; Table S5: miRNA statistical results of each sample; Table S6: miRNA expression level statistics table; Table S7: miRNAs and their targets in degradome analysis; Table S8: miRNAs and their corresponding target genes of the degradome analysis.

Author Contributions: X.Z. and X.L. designed the experiments; J.C. and Z.W. conducted the experiments; X.Z., Y.H., Y.L. and Z.S. interpreted the data; J.C. and X.Z. wrote the manuscript; Y.H., W.C., Y.L. and X.L. reviewed the manuscript. All authors have read and agreed to the published version of the manuscript.

Funding: This work was supported by the Natural Science Foundation of Guangdong Province (grant no. 2021A1515012435), the Characteristic Innovation Project of the Guangdong Provincial Department of Education (grant no. 2017KTSCX017), the Guangzhou Science and Technology Project Scientific Special (No201904010011), the National Natural Science Foundation of China (grant no. 31701970 and 31372112), and the Pearl River Talent Program for Young Talent (grant no. 2017GC010321).

Data Availability Statement: The data presented in this study are available on request from the corresponding author.

Conflicts of Interest: The authors declare that the research was conducted in the absence of any commercial or financial relationships that could be construed as a potential conflict of interest.

References

1. Fabi, J.P.; Mendes, L.R.B.C.; Lajolo, F.M.; do Nascimento, J.R.O. Transcript profiling of papaya fruit reveals differentially expressed genes associated with fruit ripening. *Plant Sci.* **2010**, *179*, 225–233. [CrossRef]
2. Girón-Ramírez, A.; Peña-Rodríguez, L.M.; Escalante-Erosa, F.; Fuentes, G.; Santamaría, J.M. Identification of the SHINE clade of AP2/ERF domain transcription factors genes in *Carica papaya*; Their gene expression and their possible role in wax accumulation and water deficit stress tolerance in a wild and a commercial papaya genotypes. *Environ. Exp. Bot.* **2021**, *183*. [CrossRef]
3. Krishna, K.L.; Paridhavi, M.; Patel, J.A. Review on nutritional, medicinal and pharmacological properties of papaya (*Carica papaya* linn.). *Indian J. Nat. Prod. Resour.* **2008**, *7*, 364–373. [CrossRef]
4. Li, X.; Zhu, X.; Mao, J.; Zou, Y.; Fu, D.; Chen, W.; Lu, W. Isolation and characterization of ethylene response factor family genes during development, ethylene regulation and stress treatments in papaya fruit. *Plant Physiol. Biochem.* **2013**, *70*, 81–92. [CrossRef] [PubMed]
5. Maringgal, B.; Hashim, N.; Tawakkal, I.S.M.A.; Mohamed, M.T.M.; Hamzah, M.H.; Ali, M.M. Effect of *Kelulut* honey nanoparticles coating on the changes of respiration rate, ascorbic acid, and total phenolic content of papaya (*Carica papaya* L.) during cold storage. *Foods* **2021**, *10*, 432. [CrossRef] [PubMed]
6. Chen, N.J.; Paull, R.E. Endoxylanase expressed during papaya fruit ripening: Purification, cloning and characterization. *Funct. Plant Biol.* **2003**, *30*, 433–441. [CrossRef] [PubMed]
7. Shiga, T.M.; Fabi, J.P.; Do Nascimento, J.R.O.; De Petkowicz, C.L.O.; Vriesmann, L.C.; Lajolo, F.M.; Cordenunsi, B.R. Changes in cell wall composition associated to the softening of ripening papaya: Evidence of extensive solubilization of large molecular mass galactouronides. *J. Agric. Food Chem.* **2009**, *57*, 7064–7071. [CrossRef]
8. Blankenship, S.M.; Dole, J.M. 1-Methylcyclopropene: A review. *Postharvest Biol. Technol.* **2003**, *28*, 1–25. [CrossRef]
9. Thewes, F.R.; Anese, R.O.; Thewes, F.R.; Ludwig, V.; Klein, B.; Wagner, R.; Nora, F.R.; Rombaldi, C.V.; Brackmann, A. Dynamic controlled atmosphere (DCA) and 1-MCP: Impact on volatile esters synthesis and overall quality of 'Galaxy' apples. *Food Packag. Shelf Life* **2020**, *26*. [CrossRef]
10. Zhu, X.; Song, Z.; Li, Q.; Li, J.; Chen, W.; Li, X. Physiological and transcriptomic analysis reveals the roles of 1-MCP in the ripening and fruit aroma quality of banana fruit (Fenjiao). *Food Res. Int.* **2020**, *130*. [CrossRef]
11. Qian, C.; Ji, Z.; Zhu, Q.; Qi, X.; Li, Q.; Yin, J.; Liu, J.; Kan, J.; Zhang, M.; Jin, C.; et al. Effects of 1-mcp on proline, polyamine, and nitric oxide metabolism in postharvest peach fruit under chilling stress. *Hortic. Plant J.* **2021**, *7*, 188–196. [CrossRef]
12. Du, Y.; Jin, T.; Zhao, H.; Han, C.; Sun, F.; Chen, Q.; Yue, F.; Luo, Z.; Fu, M. Synergistic inhibitory effect of 1-methylcyclopropene (1-MCP) and chlorine dioxide (ClO$_2$) treatment on chlorophyll degradation of green pepper fruit during storage. *Postharvest Biol. Technol.* **2021**, *171*, 111363. [CrossRef]
13. Ding, X.; Zhu, X.; Ye, L.; Xiao, S.; Wu, Z.; Chen, W.; Li, X. The interaction of CpEBF1 with CpMADSs is involved in cell wall degradation during papaya fruit ripening. *Hortic. Res.* **2019**, *6*, 1–18. [CrossRef]

14. Zheng, S.; Hao, Y.; Fan, S.; Cai, J.; Chen, W.; Li, X.; Zhu, X. Metabolomic and transcriptomic profiling provide novel insights into fruit ripening and ripening disorder caused by 1-MCP treatments in papaya. *Int. J. Mol. Sci.* **2021**, *22*, 916. [CrossRef]

15. Leticia, S.; Meza, R.; Tobaruela, E.D.C.; Pascoal, G.B.; Massaretto, I.L.; Purgatto, E. Post-harvest treatment with methyl jasmonate impacts lipid metabolism in tomato pericarp (*Solanum lycopersicum* L. Cv. Grape) at different ripening stages. *Foods* **2021**, *10*, 877. [CrossRef]

16. Park, M.-H.; Kim, S.-J.; Lee, J.-S.; Hong, Y.-P.; Chae, S.-H.; Ku, K.-M. Carbon dioxide pretreatment and cold storage synergistically delay tomato ripening through transcriptional change in ethylene-related genes and respiration-related metabolism. *Foods* **2021**, *10*, 744. [CrossRef]

17. Luo, J.; Zhou, J.J.; Zhang, J.Z. Aux/IAA gene family in plants: Molecular structure, regulation, and function. *Int. J. Mol. Sci.* **2018**, *19*, 259. [CrossRef] [PubMed]

18. He, Y.; Xue, J.; Li, H.; Han, S.; Jiao, J.; Rao, J. Ethylene response factors regulate ethylene biosynthesis and cell wall modification in persimmon (*Diospyros kaki* L.) fruit during ripening. *Postharvest Biol. Technol.* **2020**, *168*, 111255. [CrossRef]

19. Fu, C.C.; Chen, H.J.; Gao, H.Y.; Wang, S.L.; Wang, N.; Jin, J.C.; Lu, Y.; Yu, Z.L.; Ma, Q.; Han, Y.C. Papaya CpMADS4 and CpNAC3 co-operatively regulate ethylene signal genes *CpERF9* and *CpEIL5* during fruit ripening. *Postharvest Biol. Technol.* **2021**, *175*, 111485. [CrossRef]

20. Wu, Q.; Tao, X.; Ai, X.; Luo, Z.; Mao, L.; Ying, T.; Li, L. Effect of exogenous auxin on aroma volatiles of cherry tomato (*Solanum lycopersicum* L.) fruit during postharvest ripening. *Postharvest Biol. Technol.* **2018**, *146*, 108–116. [CrossRef]

21. Khaksar, G.; Sirikantaramas, S. Auxin Response Factor 2A is part of the regulatory network mediating fruit ripening through auxin-ethylene crosstalk in durian. *Front. Plant Sci.* **2020**, *11*. [CrossRef]

22. Abreu, P.M.V.; Gaspar, C.G.; Buss, D.S.; Ventura, J.A.; Ferreira, P.C.G.; Fernandes, P.M.B. Carica papaya microRNAs are responsive to *Papaya meleira virus* infection. *PLoS ONE* **2014**, *9*, e103401. [CrossRef]

23. Ding, Y.; Wang, J.; Lei, M.; Li, Z.; Jing, Y.; Hu, H.; Zhu, S.; Xu, L. Small RNA sequencing revealed various microRNAs involved in ethylene-triggered flowering process in *Aechmea fasciata*. *Sci. Rep.* **2020**, *10*, 1–10. [CrossRef]

24. Li, A.; Liu, D.; Wu, J.; Zhao, X.; Hao, M.; Geng, S.; Yan, J.; Jiang, X.; Zhang, L.; Wu, J.; et al. mRNA and small RNA transcriptomes reveal insights into dynamic homoeolog regulation of allopolyploid heterosis in nascent hexaploid wheat. *Plant Cell* **2014**, *26*, 1878–1900. [CrossRef]

25. Pagano, L.; Rossi, R.; Paesano, L.; Marmiroli, N.; Marmiroli, M. miRNA regulation and stress adaptation in plants. *Environ. Exp. Bot.* **2021**, *184*, 104369. [CrossRef]

26. Das, B.; Sen, A.; Roy, S.; Banerjee, O.; Bhattacharya, S. miRNAs: Tiny super-soldiers shaping the life of rice plants for facing "stress"-ful times. *Plant Gene* **2021**, *26*, 100281. [CrossRef]

27. Li, K.; Liu, Z.; Xing, L.; Wei, Y.; Mao, J.; Meng, Y.; Bao, L.; Han, M.; Zhao, C.; Zhang, D. miRNAs associated with auxin signaling, stress response, and cellular activities mediate adventitious root formation in apple rootstocks. *Plant Physiol. Biochem.* **2019**, *139*, 66–81. [CrossRef]

28. Zhang, T.; Li, W.; Xie, R.; Xu, L.; Zhou, Y.; Li, H.; Yuan, C.; Zheng, X.; Xiao, L.; Liu, K. CpARF2 and CpEIL1 interact to mediate auxin–ethylene interaction and regulate fruit ripening in papaya. *Plant J.* **2020**, *103*, 1318–1337. [CrossRef] [PubMed]

29. Gidhi, A.; Kumar, M.; Mukhopadhyay, K. The auxin response factor gene family in wheat (*Triticum aestivum* L.): Genome-wide identification, characterization and expression analyses in response to leaf rust. *South Afr. J. Bot.* **2020**, *000*. [CrossRef]

30. Dan, M.; Huang, M.; Liao, F.; Qin, R.; Liang, X.; Zhang, E.; Huang, M.; Huang, Z.; He, Q. Identification of ethylene responsive mirnas and their targets from newly harvested banana fruits using high-throughput sequencing. *J. Agric. Food Chem.* **2018**, *66*, 10628–10639. [CrossRef] [PubMed]

31. Zhu, X.; Ye, L.; Ding, X.; Gao, Q.; Xiao, S.; Tan, Q.; Huang, J.; Chen, W.; Li, X. Transcriptomic analysis reveals key factors in fruit ripening and rubbery texture caused by 1-MCP in papaya. *BMC Plant Biol.* **2019**, *19*, 1–22. [CrossRef]

32. Liu, Y.; Wang, L.; Chen, D.; Wu, X.; Huang, D.; Chen, L.; Li, L.; Deng, X.; Xu, Q. Genome-wide comparison of microRNAs and their targeted transcripts among leaf, flower and fruit of sweet orange. *BMC Genom.* **2014**, *15*, 1–15. [CrossRef]

33. Zhu, X.; Li, X.; Chen, W.; Chen, J.; Lu, W.; Chen, L.; Fu, D. Evaluation of new reference genes in papaya for accurate transcript normalization under different experimental conditions. *PLoS ONE* **2012**, *7*, e44405. [CrossRef]

34. Xu, Y.; Xu, H.; Wall, M.M.; Yang, J. Roles of transcription factor SQUAMOSA promoter binding protein-like gene family in papaya (*Carica papaya*) development and ripening. *Genomics* **2020**, *112*, 2734–2747. [CrossRef]

35. Wang, R.; Fang, Y.N.; Wu, X.M.; Qing, M.; Li, C.C.; Xie, K.D.; Deng, X.X.; Guo, W.W. The miR399-CsUBC24 module regulates reproductive development and male fertility in citrus. *Plant Physiol.* **2020**, *183*, 1681–1695. [CrossRef] [PubMed]

36. Aryal, R.; Yang, X.; Yu, Q.; Sunkar, R.; Li, L.; Ming, R. Asymmetric purine-pyrimidine distribution in cellular small RNA population of papaya. *BMC Genom.* **2012**, *13*. [CrossRef] [PubMed]

37. Jiang, C.M.; Wu, M.C.; Wu, C.L.; Chang, H.M. Pectinesterase and polygalacturonase activities and textural properties of rubbery papaya (*Carica papaya* Linn.). *J. Food Sci.* **2003**, *68*, 1590–1594. [CrossRef]

38. Ding, X.; Zhang, L.; Hao, Y.; Xiao, S.; Wu, Z.; Chen, W.; Li, X.; Zhu, X. Genome-wide identification and expression analyses of the calmodulin and calmodulin-like proteins reveal their involvement in stress response and fruit ripening in papaya. *Postharvest Biol. Technol.* **2018**, *143*, 13–27. [CrossRef]

39. Façanha, R.V.; Spricigo, P.C.; Purgatto, E.; Jacomino, A.P. Combined application of ethylene and 1-methylcyclopropene on ripening and volatile compound production of "Golden" papaya. *Postharvest Biol. Technol.* **2019**, *151*, 160–169. [CrossRef]

40. Wang, Y.; Wang, Q.; Gao, L.; Zhu, B.; Ju, Z.; Luo, Y.; Zuo, J. Parsing the regulatory network between small RNAs and target genes in ethylene pathway in tomato. *Front. Plant Sci.* **2017**, *8*, 1–13. [CrossRef] [PubMed]
41. Yu, Y.; Sun, F.; Chen, N.; Sun, G.; Wang, C.Y.; Wu, D.X. MiR396 regulatory network and its expression during grain development in wheat. *Protoplasma* **2021**, *258*, 103–113. [CrossRef] [PubMed]
42. Hou, G.; Du, C.; Gao, H.; Liu, S.; Sun, W.; Lu, H.; Kang, J.; Xie, Y.; Ma, D.; Wang, C. Identification of microRNAs in developing wheat grain that are potentially involved in regulating grain characteristics and the response to nitrogen levels. *BMC Plant Biol.* **2020**, *20*, 1–21. [CrossRef] [PubMed]
43. Akdogan, G.; Tufekci, E.D.; Uranbey, S.; Unver, T. miRNA-based drought regulation in wheat. *Funct. Integr. Genom.* **2016**, *16*, 221–233. [CrossRef]
44. Ragupathy, R.; Ravichandran, S.; Mahdi, M.S.R.; Huang, D.; Reimer, E.; Domaratzki, M.; Cloutier, S. Deep sequencing of wheat sRNA transcriptome reveals distinct temporal expression pattern of miRNAs in response to heat, light and UV. *Sci. Rep.* **2016**, *6*, 1–15. [CrossRef]
45. Song, G.; Zhang, R.; Zhang, S.; Li, Y.; Gao, J.; Han, X.; Chen, M.; Wang, J.; Li, W.; Li, G. Response of microRNAs to cold treatment in the young spikes of common wheat. *BMC Genom.* **2017**, *18*, 212. [CrossRef]
46. Luan, Y.; Cui, J.; Li, J.; Jiang, N.; Liu, P.; Meng, J. Effective enhancement of resistance to Phytophthora infestans by overexpression of miR172a and b in *Solanum lycopersicum*. *Planta* **2018**, *247*, 127–138. [CrossRef] [PubMed]
47. Zuo, J.; Grierson, D.; Courtney, L.T.; Wang, Y.; Gao, L.; Zhao, X.; Zhu, B.; Luo, Y.; Wang, Q.; Giovannoni, J.J. Relationships between genome methylation, levels of non-coding RNAs, mRNAs and metabolites in ripening tomato fruit. *Plant J.* **2020**, *103*, 980–994. [CrossRef] [PubMed]
48. Zuo, J.; Zhu, B.; Fu, D.; Zhu, Y.; Ma, Y.; Chi, L.; Ju, Z.; Wang, Y.; Zhai, B.; Luo, Y. Sculpting the maturation, softening and ethylene pathway: The influences of microRNAs on tomato fruits. *BMC Genom.* **2012**, *13*. [CrossRef] [PubMed]
49. Shen, C.; Yue, R.; Sun, T.; Zhang, L.; Xu, L.; Tie, S.; Wang, H.; Yang, Y. Genome-wide identification and expression analysis of auxin response factor gene family in *Medicago truncatula*. *Front. Plant Sci.* **2015**, *6*, 1–13. [CrossRef]
50. Liu, K.; Yuan, C.; Feng, S.; Zhong, S.; Li, H.; Zhong, J.; Shen, C.; Liu, J. Genome-wide analysis and characterization of Aux/IAA family genes related to fruit ripening in papaya (*Carica papaya* L.). *BMC Genom.* **2017**, *18*, 1–11. [CrossRef] [PubMed]
51. Wang, L.; Liu, Z.; Qiao, M.; Xiang, F. miR393 inhibits in vitro shoot regeneration in Arabidopsis thaliana via repressing TIR1. *Plant Sci.* **2018**, *266*, 1–8. [CrossRef]
52. Mallory, A.C.; Bartel, D.P.; Bartel, B. MicroRNA-directed regulation of Arabidopsis Auxin Response Factor17 is essential for proper development and modulates expression of early auxin response genes. *Plant Cell* **2005**, *17*, 1360–1375. [CrossRef] [PubMed]
53. Azahar, I.; Ghosh, S.; Adhikari, A.; Adhikari, S.; Roy, D.; Shaw, A.K.; Singh, K.; Hossain, Z. Comparative analysis of maize root sRNA transcriptome unveils the regulatory roles of miRNAs in submergence stress response mechanism. *Environ. Exp. Bot.* **2020**, *171*, 103924. [CrossRef]
54. Jin, L.; Yarra, R.; Zhou, L.; Zhao, Z.; Cao, H. miRNAs as key regulators via targeting the phytohormone signaling pathways during somatic embryogenesis of plants. *3 Biotech* **2020**, *10*, 1–11. [CrossRef] [PubMed]

Article

Transcriptomic Analysis of *Pseudomonas aeruginosa* Response to Pine Honey via RNA Sequencing Indicates Multiple Mechanisms of Antibacterial Activity

Ioannis Kafantaris [1,†], Christina Tsadila [1,†], Marios Nikolaidis [2], Eleni Tsavea [1], Tilemachos G. Dimitriou [1], Ioannis Iliopoulos [3], Grigoris D. Amoutzias [2] and Dimitris Mossialos [1,*]

[1] Microbial Biotechnology-Molecular Bacteriology-Virology Laboratory, Department of Biochemistry & Biotechnology, University of Thessaly, Biopolis, 41500 Larissa, Greece; kafantarisioannis@gmail.com (I.K.); tsadila@uth.gr (C.T.); elenats89@hotmail.com (E.T.); tidimitr@bio.uth.gr (T.G.D.)
[2] Bioinformatics Laboratory, Department of Biochemistry & Biotechnology, University of Thessaly, Biopolis, 41500 Larissa, Greece; marionik23@gmail.com (M.N.); amoutzias@bio.uth.gr (G.D.A.)
[3] Department of Basic Sciences, School of Medicine, University of Crete, 71003 Heraklion, Greece; iliopj@med.uoc.gr
* Correspondence: mosial@bio.uth.gr; Tel.: +30-2410-565270
† These authors contributed equally.

Citation: Kafantaris, I.; Tsadila, C.; Nikolaidis, M.; Tsavea, E.; Dimitriou, T.G.; Iliopoulos, I.; Amoutzias, G.D.; Mossialos, D. Transcriptomic Analysis of *Pseudomonas aeruginosa* Response to Pine Honey via RNA Sequencing Indicates Multiple Mechanisms of Antibacterial Activity. *Foods* 2021, 10, 936. https://doi.org/10.3390/foods10050936

Academic Editors: Yelko Rodríguez-Carrasco and Bojan Šarkanj

Received: 5 April 2021
Accepted: 22 April 2021
Published: 24 April 2021

Publisher's Note: MDPI stays neutral with regard to jurisdictional claims in published maps and institutional affiliations.

Abstract: Pine honey is a unique type of honeydew honey produced exclusively in Eastern Mediterranean countries like Greece and Turkey. Although the antioxidant and anti-inflammatory properties of pine honey are well documented, few studies have investigated so far its antibacterial activity. This study investigates the antibacterial effects of pine honey against *P. aeruginosa* PA14 at the molecular level using a global transcriptome approach via RNA-sequencing. Pine honey treatment was applied at sub-inhibitory concentration and short exposure time ($0.5 \times$ of minimum inhibitory concentration –MIC- for 45 min). Pine honey induced the differential expression (>two-fold change and $p \leq 0.05$) of 463 genes, with 274 of them being down-regulated and 189 being up-regulated. Gene ontology (GO) analysis revealed that pine honey affected a wide range of biological processes (BP). The most affected down-regulated BP GO terms were oxidation-reduction process, transmembrane transport, proteolysis, signal transduction, biosynthetic process, phenazine biosynthetic process, bacterial chemotaxis, and antibiotic biosynthetic process. The up-regulated BP terms, affected by pine honey treatment, were those related to the regulation of DNA-templated transcription, siderophore transport, and phosphorylation. Pathway analysis revealed that pine honey treatment significantly affected two-component regulatory systems, ABC transporter systems, quorum sensing, bacterial chemotaxis, biofilm formation and SOS response. These data collectively indicate that multiple mechanisms of action are implicated in antibacterial activity exerted by pine honey against *P. aeruginosa*.

Keywords: *Pseudomonas aeruginosa*; pine honey; RNA-sequencing; antimicrobial activity; transcriptomics; biological process

1. Introduction

Pseudomonas aeruginosa is a ubiquitous, Gram-negative opportunistic human pathogen that can cause acute and chronic human infections in hospitalized or immune-compromised patients [1,2]. Typically, it infects the airway, urinary tract, burns, wounds, surgical site infections and also causes systemic blood infections that can lead to death [3]. The pathogenesis of *P. aeruginosa* is attributed to a variety of virulence factors, such as the cytotoxic pigment pyocyanin, the major siderophore pyoverdine, alkaline protease, elastase, exotoxins, flagella, and biofilm formation [4]. In addition, core genome analyses have revealed a distinct set of *P. aeruginosa* specific genes, related to its pathogenicity and lifestyle [5]. *P. aeruginosa* can adapt to a wide variety of environmental conditions and exhibits a remarkable high multidrug resistance by the formation of biofilms [6–8]. Considering its high prevalence associated with high mortality rates and lack of treatment options, this pathogen has been identified by the World Health Organization as a critical research priority for the development of alternative drugs and novel therapeutic strategies [9].

Recently, diverse natural products exerting antimicrobial activity have been widely investigated as alternative therapeutic agents to combat multidrug resistant pathogens. Honey, a natural product of honey bees, has been traditionally used in treating wounds and infectious diseases [10–12]. Many studies have proved the antimicrobial activity of different honey types against a plethora of pathogenic bacteria [13–17]. Previous studies conducted by our research group have also demonstrated that Greek and Cypriot honeys of diverse botanical origin exhibited potent antibacterial activity [18–20]. The antibacterial activity of honey to a wide range of pathogens is due to multiple factors including hydrogen peroxide (H_2O_2), low pH, methylglyoxal (MGO), antimicrobial peptides, and osmotic stress [16,21,22]. Several studies have shown numerous biological processes in bacteria that may be affected by honey such as protein synthesis, quorum sensing (QS), motility, biofilm formation, as well as response to oxidative stress [23–26].

Pine honey, is a unique type of honeydew honey produced in Eastern Mediterranean *Pinus brutia* and *Pinus halepensis* Miller forests, located in Greece and Turkey [27]. It is produced by bees which collect honeydew (sugary secretions) eliminated by the insect *Marchalina hellenica* (Gennadius), when feeding on certain pine trees [28]. Pine honey has an impressive pearl-amber color with characteristic metallic highlights, a spicy taste, as well as a thick texture. All the above characteristics combined with its natural property not to crystallize and its high content of minerals (potassium, calcium, iron, phosphorus, magnesium, sodium, and zinc), make pine honey a natural product of significant economic value [29]. It is estimated that in Greece and Turkey pine honey represents, 65% and 50% of the total annual honey production, respectively [30,31]. Although the antioxidant and anti-inflammatory properties of pine honey are well documented [32–35], few studies have investigated its antibacterial activity [18,36,37].

RNA sequencing (RNA-seq) is a cutting-edge technology for transcriptome profiling that can provide measurements of genome-wide quantitative analysis of all transcripts with high accuracy and sensitivity [38,39]. In addition, RNA-seq can reveal specific biological processes, affected in the presence of natural products or drugs [40,41]. To our knowledge, this is the first study that employs a global transcriptome approach via RNA-seq analysis to investigate the antibacterial effects of pine honey at the molecular level using as a model microorganism the *P. aeruginosa* PA14 strain.

2. Materials and Methods

2.1. Honey Samples

Pine honey was harvested in August 2019 from an apiary located in Chalkidiki area (Greece). After the collection, the sample was stored in glass container at room temperature in the dark. Manuka honey UMF 24+ (MGO 1122), Steens™, New Zealand (Batch No B084E3) was also used in this study.

2.2. Bacterial Strain, Growth Media, and Culture Conditions

The antibacterial activities of the pine and manuka honeys were tested upon *Pseudomonas aeruginosa* PA14 strain. The bacterial strain was routinely grown in Mueller-Hinton (MH) broth or MH agar (Lab M, Bury, UK) at 37 °C.

2.3. Assessment of the Minimum Inhibitory Concentration (MIC) and Minimum Bactericidal Concentration (MBC)

The minimum inhibitory concentration (MIC) of the tested honey sample was assessed in sterile 96-well polystyrene microtiter plate (Kisker Biotech GmbH & Co. KG, Steinfurt, Germany) using a spectrophotometric bioassay, as previously described [19]. Briefly, overnight bacterial culture grown in MH broth was adjusted to a 0.5 McFarland turbidity standard (~1.5×10^8 cfu/mL). Approximately 5×10^4 cfu in 10 µL MH broth was added to 190 μL of diluted test honey in MH broth. The control wells contained only MH broth, inoculated with bacteria. The optical density (OD) was determined at 630 nm using an EL x808 Absorbance microplate reader (BioTek Instruments, Inc., Winooski, VT, USA) before (t = 0) and after 24 h of incubation (t = 24) at 37 °C. The OD for each replicate well at t = 0 was subtracted from the OD of the same replicate well at t = 24. The growth inhibition at each honey dilution was measured using the formula: % inhibition = 1 − (OD test well/OD of corresponding control well) × 100. MIC was defined as the lowest honey concentration which results in 100% growth inhibition.

The MBC is the lowest concentration of any antibacterial agent that could kill tested bacteria. The MBC was determined by transferring a small quantity of sample contained in each replicate well of the microtiter plates to MH agar plates by using a microplate replicator (Boekel Scientific, Waltham, PA, USA). The plate was incubated at 37 °C for 24 h. The MBC was determined as the lowest honey concentration at which no grown colonies were observed [42].

2.4. Assessment of the Antibacterial Activity Attributed to Hydrogen Peroxide and Proteinaceous Compounds

The MIC of honey treated with bovine catalase or proteinase K was assessed in comparison to the untreated honey as previously described [18,43]. Briefly, 50% (*v/v*) honey in MH broth containing 100 mg/mL proteinase K (Blirt, Gdansk, Poland) or 600 U/mL bovine catalase (Serva, Heidelberg, Germany) was incubated for 16 h at 37 °C, then diluted and tested as described above. An increased MIC of treated honey compared to the untreated honey has shown that the antibacterial activity of tested honey was attributable to hydrogen peroxide and/or proteinaceous compounds, respectively.

2.5. Total RNA Isolation and RNA Sequencing

Pseudomonas aeruginosa PA14 culture was prepared in MH broth to an initial optical density at 600 nm (OD_{600}) of 0.05 and then incubated in a 250-mL cell culture conical flask (Erlenmeyer, Duran) at 37 °C with shaking at 200 rpm until reaching mid-exponential phase (OD_{600} of 0.4). Cultures were then split into two conical sterile falcons (Falcon, Corning): one falcon contained 30 mL of untreated culture (control) and the second falcon (30 mL) contained the culture and the treatment at a final concentration of roughly 0.5× MIC of pine honey (4.5% *v/v*). Each culture was grown for an additional 45 min before total RNA isolation. The control and the treated culture were then split into three technical replicates (10 mL each). Total RNA from each replicate was isolated using a NucleoSpin RNA isolation kit (Macherey-Nagel) and DNA removed with DNase I, according to the manufacturer's protocol. Samples were analyzed spectrophotometrically using a micro-volume UV-Vis instrument (Quawell, San Jose, CA, USA) for quantification and purity assessment. All RNA samples had an $A_{260}{:}A_{280}$ ratio between 1.8 and 2.0. RNA integrity was initially verified by 1% agarose gel electrophoresis. The six samples (3 controls and 3 treated) were shipped to Macrogen (Seoul, South Korea) for rRNA depletion using a NEBNext Bacterial rRNA removal kit (Illumina, San Diego, CA, USA), library preparation using the TruSeq stranded total RNA kit (Illumina, San Diego, CA, USA), and subsequent 150-bp paired-end RNA sequencing on a NovaSeq6000 platform (Illumina, San Diego, CA, USA). RNA integrity was further evaluated using the RNA Nano 6000 Assay Kit of the Agilent Bioanalyzer 2100 system (Agilent Technologies, Santa Clara, CA, USA); all samples demonstrated RNA integrity number (RIN) > 8.0.

2.6. Bioinformatics Analysis of the Differentially Expressed Genes (DEGs)

The fastq files were downloaded from Macrogen, with adapter trimming applied (TruSeq3 paired-ended) and their read quality was initially assessed using FASTQC (version 0.11.5) (http://www.bioinformatics.babraham.ac.uk/projects/fastqc, accessed on 20 December 2020). Subsequently, the reads were trimmed with Trimmomatic (version 0.38-default parameters) [44] and their quality was again assessed. Reads were aligned to the *P. aeruginosa* PA14 genome (NCBI reference sequence, NC_008463.1; GenBank accession number CP000438.1; assembly GCA_000014625.1) using the alignment program HISAT2 [45] and subsequently the number of reads that mapped to each gene was counted using the feature counts tool (version 1.6.4) with default parameters [46]. After mapping and counting, differential expression analysis, between control and treated samples, was carried out using the DESeq2 package (version 2.11.39) [47]. Gene annotation was carried out using the tool of DESeq2 package and the appropriate file (gtf) from the assembly (GCA_000014625.1). Genes whose expression displayed an average fold change >2 and was statistically significant (adjusted p value ≤ 0.05) were considered differentially expressed (DEGs). In order to understand more profoundly the biological functions and the metabolic pathways of the identified genes, the DEGs were functionally classified due to Gene Ontology (GO), using the Goseq tool (version 3.12) [48] and Kyoto Encyclopedia of Genes and Genomes (KEGG) database (http://www.genome.jp/kegg/, accessed on 27 December 2020) [49]. Go annotation and KEGG classifications were downloaded from the Pseudomonas Community Annotation Project (PseudoCAP) [50].

A second RNA sequencing analysis using a different pipeline was also conducted for assessing the robustness of the RNA-seq analysis conclusions. Briefly, the fastq files were trimmed with minimum contig length parameters and the quality of the final reads was inspected with FastQC. The trimmed fastq files were used for the de novo assembly of the *P. aeruginosa* PA14 transcriptome with the Trinity software (default parameters) [51], using all samples together. The resulting 5674 Trinity contigs were filtered to keep the longest isoform of each trinity gene, thus retaining 4903 contigs. The 4903 contigs were fed to TransDecoder software (https://github.com/TransDecoder/TransDecoder, accessed on 15 January 2021) with default parameters to identify the putative ORFs. Next the resulting 15,549 TransDecoder CDS were used as database for BLASTn search while the *P. aeruginosa* PA14 (GCA_000014625.1) CDS were used as a query with e-value cut-off 1×10^{-5}. The best BLAST hit was kept for each *P. aeruginosa* PA14 gene, resulting in 4614 TransDecoder CDS. Next, Bowtie2 [52] within Trinity was used to align the reads back to the 4,614 TransDecoder CDS. In addition to the Trinity de novo assembly, a reference guided analysis was also performed by aligning the trimmed reads to the *P. aeruginosa* PA14 CDS with Bowtie2. Identification of DEGs in both analyses was conducted with the edgeR package [53] within Trinity, using default parameters.

3. Results and Discussion

3.1. Antibacterial Activity of Pine Honey against P. aeruginosa PA14

In order to investigate the activity of pine honey against *P. aeruginosa*, MIC and MBC values were determined. Data are presented in Table 1.

Table 1. Antibacterial activity of pine honey and manuka against *P. aeruginosa*.

Honey	MIC % (*v/v*) [1]	MBC % (*v/v*) [2]	MICp % (*v/v*) [3]	MICc % (*v/v*) [4]
pine honey	9	9	9	20
manuka	9	11	ND [5]	ND

[1] MIC, minimum inhibitory concentration. [2] MBC, minimum bactericidal concentration. [3] MICp, MIC values of proteinase K treated honey. [4] MICc, MIC values of catalase treated honey. [5] ND, not determined.

The results clearly demonstrate that pine honey and manuka exert high anti-bacterial activity since both inhibited *P. aeruginosa* at 9% (*v/v*). Furthermore, pine honey was bactericidal at 9% (*v/v*), while manuka was bactericidal at 11% (*v/v*). In order to investigate the mechanisms which may contribute to the anti-bacterial activity, pine honey was treated with catalase and proteinase K. The proteinase K treated pine honey exhibited MIC value 9% (*v/v*) against *P. aeruginosa*, which is the same as the untreated honey, while the catalase treated honey exhibited higher MIC value 20% (*v/v*) indicating that the anti-bacterial activity of the pine honey was mainly attributable to hydrogen peroxide and not to proteinaceous compounds (Table 1). On the other hand, it is known that MGO is the main antimicrobial compound in manuka honey [22].

3.2. Effects of Pine Honey on the Transcriptomic Profile of P. aeruginosa PA14

3.2.1. Global Response of *P. aeruginosa* to Pine Honey Treatment

The molecular response of *P. aeruginosa* to pine honey was investigated using RNA-Seq. Pine honey treatment was applied at sub-inhibitory concentration and short exposure time ($0.5\times$ MIC for 45 min), since this approach induces more specific response and reduces indirect effects [54]. A sub-inhibitory concentration may act as stress inducer or cues/coercion on receiver bacteria [55].

Data analysis of DEGs was conducted using three different pipelines, leading to very similar results and conclusions. RNA-seq analysis, using the pipeline (HISAT2-featurecounts-DESeq2), revealed that pine honey significantly affects the transcriptomic profile of *P. aeruginosa* PA14 compared to the control, with changes to the expression of 2543 out of 5964 coding sequences (42.6%; $p \leq 0.05$). Of those 2543 genes, 1257 were up-regulated (21% of all coding genes) and 1286 were down-regulated (21.6%). The second pipeline (de novo Trinity-edgeR) revealed that pine honey induced the differential expression of 2115 out of 4673 genes (45.2%; $p \leq 0.05$), where 1112 were up-regulated (23.8% of all coding genes) and 1103 were down-regulated (23.6%). In addition, the third pipeline (reference genome guided analysis-Bowtie2-edgeR) showed that pine honey induced the differential expression of 2451 out of 5964 coding sequences (41.1%; $p \leq 0.05$) where 1195 were up-regulated (20% of all coding genes) and 1256 were down-regulated (21.1%). In a similar study, treatment of *P. aeruginosa* PA14 with manuka honey induced the differential expression of 3177 genes (54%; $p \leq 0.05$) with 1646 of them being up-regulated (representing 28% of all coding genes) and 1531 being down-regulated (26% of all coding genes) [56]. Genome-wide expression changes were visualized as heatmap and volcano plot to identify specific genes with high fold changes and statistical significance. Results of hierarchical clustering and volcano plot are shown in Figures 1A and 2.

The results using the first pipeline (HISAT2-featurecounts-DESeq2) showed that pine honey treatment strongly induced the differential expression ($\log_2$FC > 1, meaning >two-fold change and $p \leq 0.05$) of 463 genes (7.8% of all coding sequences) including 274 down-regulated and 189 up-regulated genes (Table S1). The results of the second pipeline (de novo Trinity-edgeR), revealed that pine honey treatment induced the differential expression ($\log_2$FC > 1 and $p \leq 0.05$) of 440 genes (9.4% of all coding sequences) including 265 down-regulated and 175 up-regulated genes (Table S2). In addition, the last pipeline (reference guided analysis-Bowtie2-edgeR) showed that pine honey induced the differential expression ($\log_2$FC > 1 and $p \leq 0.05$) of 482 genes (8.1% of all coding sequences) including 192 up-regulated and 290 down-regulated genes (Table S3). Further data analysis regarding DEGs was conducted using the pipeline (HISAT2-featurecounts-DESeq2). Compared to the study conducted by Bouzo et al. [56], treatment of *P. aeruginosa* PA14 with manuka honey highly induced the differential expression of 235 genes ($\log_2$FC > 2 meaning >four-fold change and $p \leq 0.05$) including more up-regulated than down-regulated genes. In Figure 2, genes that were significantly differentially expressed are presented in red (up-regulated) and blue color (down-regulated). The most up- and down-regulated genes are labeled in each plot. In addition, Figure 1B shows the bi-plot of the principal-component analysis of DESeq2 normalized read counts (all coding genes) for pine honey treatment (green) and the control (red), split into technical replicates. Principal component analysis (PCA) confirmed that the effect of pine honey on *P. aeruginosa* differed significantly relative to the control (Figure 1B).

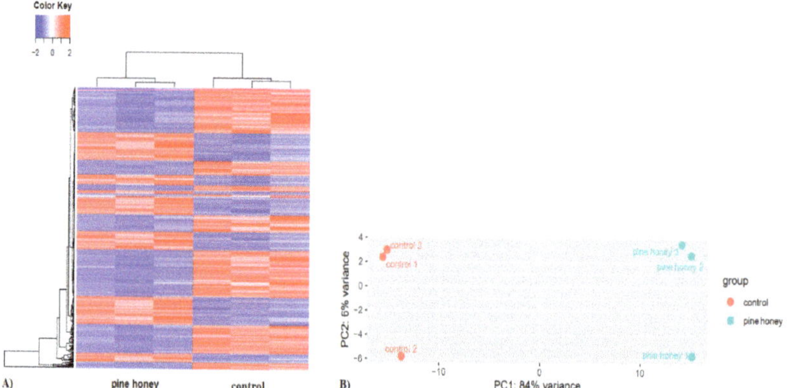

Figure 1. Transcriptional response of *P. aeruginosa* PA14 treated at mid-exponential phase with pine honey for 45 min at 0.5× MIC. (**A**) Clustered heatmap (based on Euclidean measures and complete agglomeration) of all DEGs (>two fold changes and *p* value ≤ 0.05) in *P. aeruginosa* across pine honey treatment and control. Each column represents one sample, and each row represents one gene. The red and blue gradients indicate up- and down-regulated gene expression, respectively. (**B**) Bi-plot of the principal-component analysis of DESeq2 normalized read counts (all coding genes) for pine honey treatment (green) and the control (red), split into technical replicates.

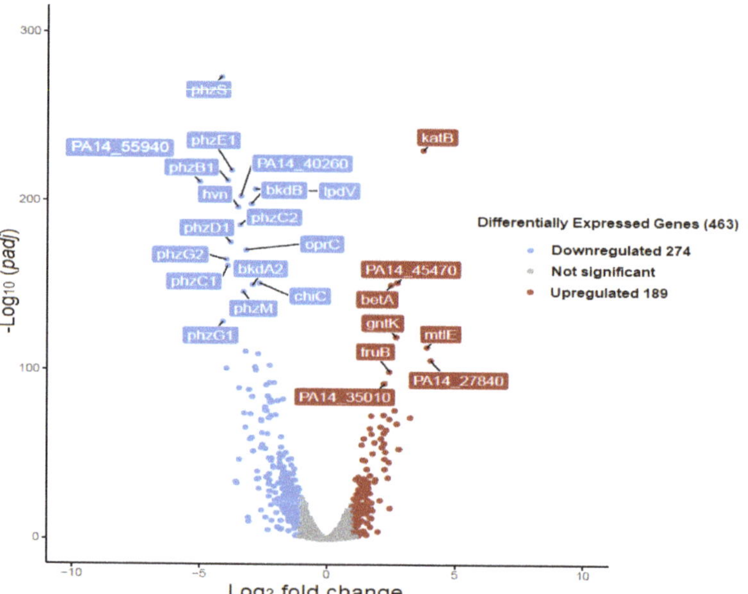

Figure 2. Volcano plot of differentially expressed genes (DEGs) based on RNA-seq analysis of untreated and pine honey-treated *Pseudomonas aeruginosa* PA14. Each gene is represented by a dot in the graph and the most differentially expressed up-and down-regulated genes are labeled in each plot. The x-axis and y-axis represent the $\log_2$ value of the fold change and the t-statistic as $-\log_{10}$ of the *p*-value, respectively. The genes represented in red (up-regulated) and blue (down-regulated) are differentially expressed genes with >two fold changes and a *p* value ≤ 0.05, while gray dots show genes with no significant difference compared to the control.

3.2.2. Top Up- and Down-Regulated DEGs

In the pine honey-treated samples, the genes *katB*, PA14_45470, *betA*, *gntK*, *mtlE*, *fruB*, PA14_27840, and PA14_35010 were among the top up-regulated. These genes encode the catalase enzyme *katB* ($\log_2$FC 3.72), a putative glutathione S-transferase, the choline dehydrogenase *betA*, a gluconokinase, a putative binding protein component of ABC maltose/mannitol transporter ($\log_2$FC 3.89), a putative phosphotransferase system fructose-specific component, a putative copper-binding protein ($\log_2$FC 4.01) and a hypothetical protein respectively (Figure 2). It is documented that the catalase enzyme *katB* and glutathione S-transferases are induced in the presence of hydrogen peroxide. Moreover, these enzymes play multiple crucial roles in oxidative stress protection and bacterial virulence in *P. aeruginosa* [57–59]. Another enzyme, the choline dehydrogenase *betA* contributes toward the hyperosmotic stress resistance in *Pseudomonas protegens* [60]. Therefore the observed transcriptional response clearly shows that *P. aeruginosa* cells attempt to adapt to the hostile environment of pine honey, which is characterized by the presence of hydrogen peroxide and high osmolarity.

Among the genes that were strongly down-regulated were those encoding proteins involved in phenazine biosynthesis *phzB*1, *C*1, *C*2, *D*1, *E*1 ($\log_2$FC ranged from −3.40 to −3.90), PA14_55940 ($\log_2$FC −5.20) and PA14_40260 ($\log_2$FC −3.39) encoding a putative pilus assembly protein and a conserved hypothetical protein, respectively. Interestingly, KEEG pathway analysis (see also further below) revealed that PA14_40260 encodes a protein involved in the pathway of quorum sensing whereas, curated search in both KEEG and PseudoCAP databases revealed that PA14_55940, the most down-regulated gene in the presence of pine honey, encodes a bacterial motility protein (fimbriae associated protein Flp/Fap pilin component) of the protein secretion/export apparatus (Type II secretion system) [49,50]. Our observations are in accordance with a relevant study, where manuka honey reduced the motility of *P. aeruginosa* through the suppression of flagellin-associated genes [25]. It is plausible that pine honey reduces in a similar way the motility thus reducing *P. aeruginosa* virulence.

Other genes that were also strongly down-regulated include *phzS* ($\log_2$FC −4.16) and M ($\log_2$FC −3.28) encoding a flavin-containing monooxygenase and a probable phenazine-specific methyltransferase respectively, *oprC* ($\log_2$FC −3.17) encoding an outer membrane copper receptor (pores ion channels), *hvn* ($\log_2$FC −3.50) encoding a putative halovibrin protein, *bkdB* encoding a lipoamide acyltransferase component of branched-chain alpha-keto acid dehydrogenase complex E2, *lpdV* (lipoamide dehydrogenase-Val) and *chiC* that encode a chitinase (Figure 2).

3.2.3. Gene Ontology (GO) Enrichment Analysis

In order to further investigate the biological functions and the metabolic pathways of DEGs in presence of pine honey, GO analysis was performed [50,61]. The most enriched GO categories among the DEGs are shown in Figure 3 and Supplementary Tables S4–S7.

In the Biological Processes (BP) category (total DEGs: 375, up-regulated: 177, down-regulated: 198), the most enriched terms for up-regulated DEGs in presence of pine honey were related to "regulation of DNA-templated transcription," "siderophore transport," and "phosphorylation" whereas, in contrast, the most enriched BP GO terms for down-regulated DEGs were "oxidation-reduction process," "transmembrane transport," "proteolysis," "signal transduction," "biosynthetic process," "phenazine biosynthetic process," "bacterial chemotaxis," and "antibiotic biosynthetic process" (Figure 3A, Table S5). In the Cellular Component (CC) category (total DEGs: 138, up-regulated: 62, down-regulated: 76), the most enriched terms for up-regulated DEGs in presence of pine honey were related to "cell outer membrane" and "integral component of plasma membrane" whereas the most enriched CC GO terms for down-regulated DEGs were "integral component of membrane," "ATP-binding cassette (ABC) transporter complex," and "cytoplasm" (Figure 3B, Table S6). In addition, in this category, the most enriched GO term was "membrane." Furthermore, in the Molecular Function (MF) category (total DEGs: 513, up-regulated: 233, down-regulated:

280) the most enriched terms for up-regulated DEGs in presence of pine honey were "DNA and ATP binding" whereas, "catalytic activity" and "flavin adenine dinucleotide binding" were the most enriched terms for down-regulated DEGs. Other highly enriched MF GO terms were "oxidoreductase activity" and "transmembrane transporter activity" (Figure 3C, Table S7).

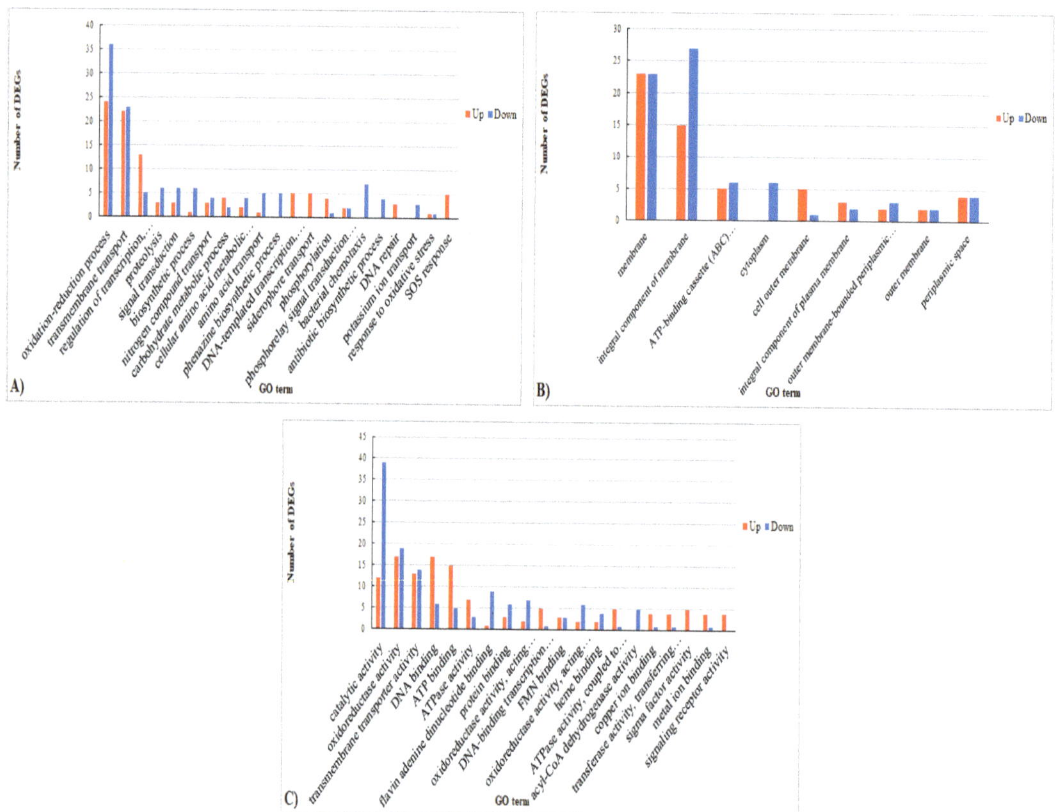

Figure 3. Gene Ontology categories of the DEGs. Three main categories were identified: (**A**) biological process (BP, total DEGs: 375, up-regulated: 177, down-regulated: 198) (**B**) cellular component (CC, total DEGs: 138, up-regulated: 62, down-regulated: 76) (**C**) molecular function (MF, total DEGs: 513, up-regulated: 233, down-regulated: 280).

3.2.4. KEGG Pathway Enrichment Analysis

Kyoto Encyclopedia of Genes and Genomes (KEGG) pathway enrichment analysis revealed that pine honey significantly affected several cellular pathways and induced the differential expression of genes involved in (but not limited to) two-component regulatory systems, ABC transporters, quorum sensing (QS), bacterial chemotaxis, and biofilm formation. Regarding the two-component regulatory systems, pine honey treatment caused significant up-regulation of 10 genes and down-regulation of 25 genes (Figure 4A).

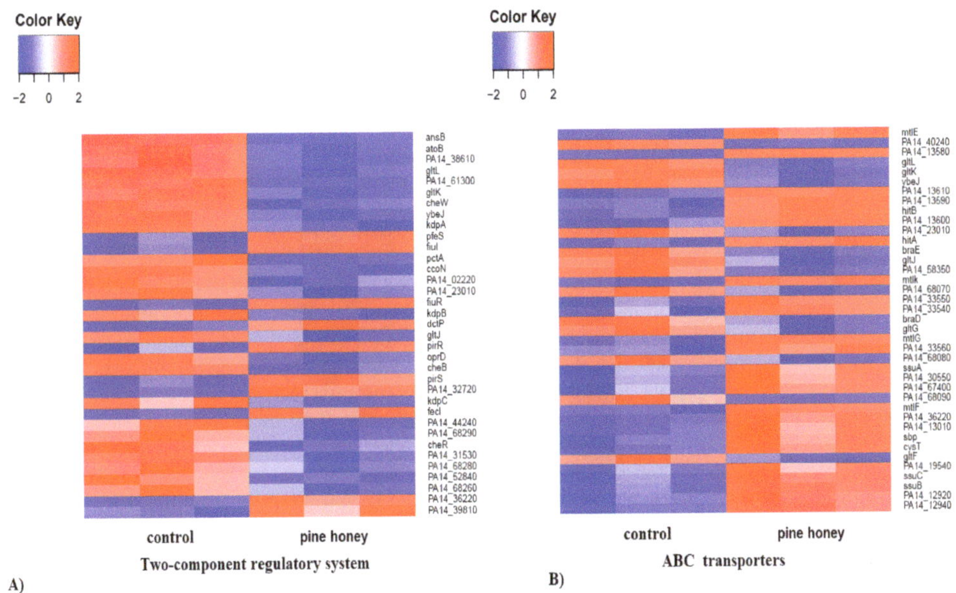

Figure 4. Heatmaps show log$_2$FC data of DEGs implicated in: (**A**) two-component regulatory system. (**B**) ABC transporters, according to KEGG analysis. The red and blue gradients indicate up- and down-regulated gene expression.

In treated samples, two genes (*pfeS* and *pirR*) encoding a sensor and response regulator respectively, were among the most up-regulated (log$_2$FC 1.61 and 1.58, respectively). In contrast, the most down-regulated genes were *atoB*, PA14_38610, *ansB*, PA14_31530, *kdpA, B* and *C* (log$_2$FC ranged from −1.62 to −3.22). These genes encode an acetyl-CoA acetyltransferase, a putative short-chain fatty acid transporter, a glutaminase-asparaginase, a putative acyl-CoA thiolase, and potassium-transporting ATPase chain ABC, respectively. Other down-regulated DEGs include *oprD* encoding an outer membrane porin and *cheW*, encoding a putative purine-binding chemotaxis protein. In the ABC transporter gene group the up-regulated were more prevalent than the down-regulated genes (Figure 4B). The most up-regulated genes were *mtlE, K, G* (log$_2$FC ranged from 1.56 to 3.89) encoding putative ATP-binding component of ABC maltose/mannitol transporters whereas, the most down-regulated genes were PA14_40240 and *gltK, L* encoding a putative ATP-binding/permease fusion and, putative permease and ATP-binding component of ABC transporter system respectively. Interestingly, apart from the down-regulated genes *oprC* and *D*, encoding outer membrane porins, *oprB* (log$_2$FC −1.65) encoding a glucose/carbohydrate outer membrane porin, and PA14_58410 (log$_2$FC −1.55) encoding putative membrane porin, were also down-regulated. In contrast, the genes *mexF* and *E*, encoding putative RND efflux transporter and RND efflux membrane fusion protein precursor, were up-regulated (log$_2$FC, 2.25 and 2.42, respectively). *OprC* is a porin abundant in the outer membrane vesicles involved in channel-forming and copper binding [62]. *OprC* transports copper, an essential trace element implicated in several physiological processes, into bacteria during copper deficiency. In a very recent study the authors showed that *oprC* deletion inhibited bacterial motility and quorum-sensing systems, as well as decreased lipopolysaccharide and pyocyanin levels in *P. aeruginosa* [62]. Interestingly, a previous study has shown that manuka honey decreased pyocyanin production in *P. aeruginosa* PA14, presumably via interaction with the MvfR quorum sensing network [63]. The *oprD* porin facilitates the diffusion of basic amino acids and peptides containing these residues. Moreover, it is implicated in carbapenem resistance [64]. On the other hand, the *oprB* porin has been associated with the diffusion of glucose across the outer membrane of *P. aeruginosa* thanks to the ABC transporter *glt* [65,66]. Raneri et al. [67] have demonstrated that *P. aeruginosa* mutants defective in glucose uptake have pleiotropic phenotype and attenuated virulence in non-mammal infection models. In this study, both *oprB* porin and *glt* ABC transporter were down-regulated. Previous studies have shown that reduced permeability of the outer membrane through *oprD* impairment and overexpression of the major resistance-nodulation-division (RND) efflux pump systems (MexAB-OprM, MexCD-OprJ, MexEF-OprN, and MexXY-OprM), contribute to carbapenem resistance in *P. aeruginosa* [68,69]. In this study, *oprD* is down-regulated in contrast to *mexF* and *mexE* (components of MexEF-OprN RND efflux pump system) which are up-regulated in the presence of pine honey. It is tempting to speculate that such differential gene expression might counteract the anti-bacterial activity of compounds (e.g., phytochemicals) contained in pine honey.

Furthermore, RNA-seq analysis revealed that a group of genes implicated in iron uptake and transport are up-regulated when *P. aeruginosa* PA14 is exposed to pine honey. These genes include *fptA, fecA, fpvA, piuA,* and *tonB* ($\log_2$FC ranged from 1.05 to 2.43) encoding the Fe(III)-pyochelin outer membrane receptor, a TonB-dependent siderophore receptor, the ferripyoverdine receptor, a putative outer membrane ferric siderophore receptor, and periplasmic protein TonB, respectively. Moreover, two genes *pchD* and *pchE*, implicated in pyochelin biosynthesis, were up-regulated ($\log_2$FC 1.13 and 1.14, respectively). Iron is a key nutrient, involved in many crucial biological processes. Therefore, it is essential for bacterial growth and virulence. In order to overcome restricted iron bioavailability, *P. aeruginosa* developed various strategies to acquire iron through the direct production of siderophores such as pyoverdine as well as pyochelin and the uptake of siderophores via TonB-dependent receptors (TBDRs) [70]. Several studies have shown that TBDRs could be employed in a "Trojan horse" strategy, in which the interaction between a siderophore and an antibiotic could significantly increase the antibiotic bioactivity, by facilitating its transport into the bacterial cell [71–73]. Previous reports have demonstrated the involvement of different TBDRs such as *piuA, fpvA, fecA,* and *fptA* in the uptake of siderophore-drug conjugates in *P. aeruginosa* [73,74]. Our data suggest that honey might impose an iron-limited environment for *P. aeruginosa*, which could be potentially exploited in combination with siderophore-antibiotic conjugates as an alternative approach to combat this multi-drug resistant pathogen.

Pine honey treatment significantly affected the expression of several genes involved in quorum sensing (QS), bacterial chemotaxis, and biofilm formation pathways (Figure 5A–C).

Interestingly, pine honey treatment provoked significant down-regulation of almost all genes involved in the above pathways. The values of $\log_2$FC ranged from -1.02 to -5.20. The genes *phzG1* and *G2* ($\log_2$FC -4.08 and -3.95, respectively) encode a probable pyrodoxamine 5′-phosphate oxidase whereas, the genes *lasA, B,* and *lecB* ($\log_2$FC $-1.90, -2.4,$ and -2.97 respectively) encode a staphylolytic exoprotease preproenzyme, an elastase, and a fucose-binding lectin PA-IIL, respectively. In the biofilm formation pathway the identified genes were *pa1L*, PA14_34050, PA14_34070, PA14_34100, PA14_34030, and PA14_34000 ($\log_2$FC ranged from -1.02 to -2.9) encoding a PA-I galactophilic lectin and conserved hypothetical proteins, respectively whereas, in the bacterial chemotaxis pathway the involved genes were PA14_61300, *cheW, pctA*, PA14_02220, PA14_58350, *cheB, cheR* ($\log_2$FC ranged from -1.02 to -1.67) encoding various chemotaxis proteins (i.e., methyltransferase, methyl-accepting and purine-binding).

Furthermore, pine honey induced the differential expression of genes involved in SOS response such as *lexA, recA, N, X,* and PA14_25150 ($\log_2$FC ranged from 1.2 to 1.72). Similarly, Bouzo et al. (2020) have demonstrated that manuka honey significantly up-regulated a wide range of genes involved in SOS response.

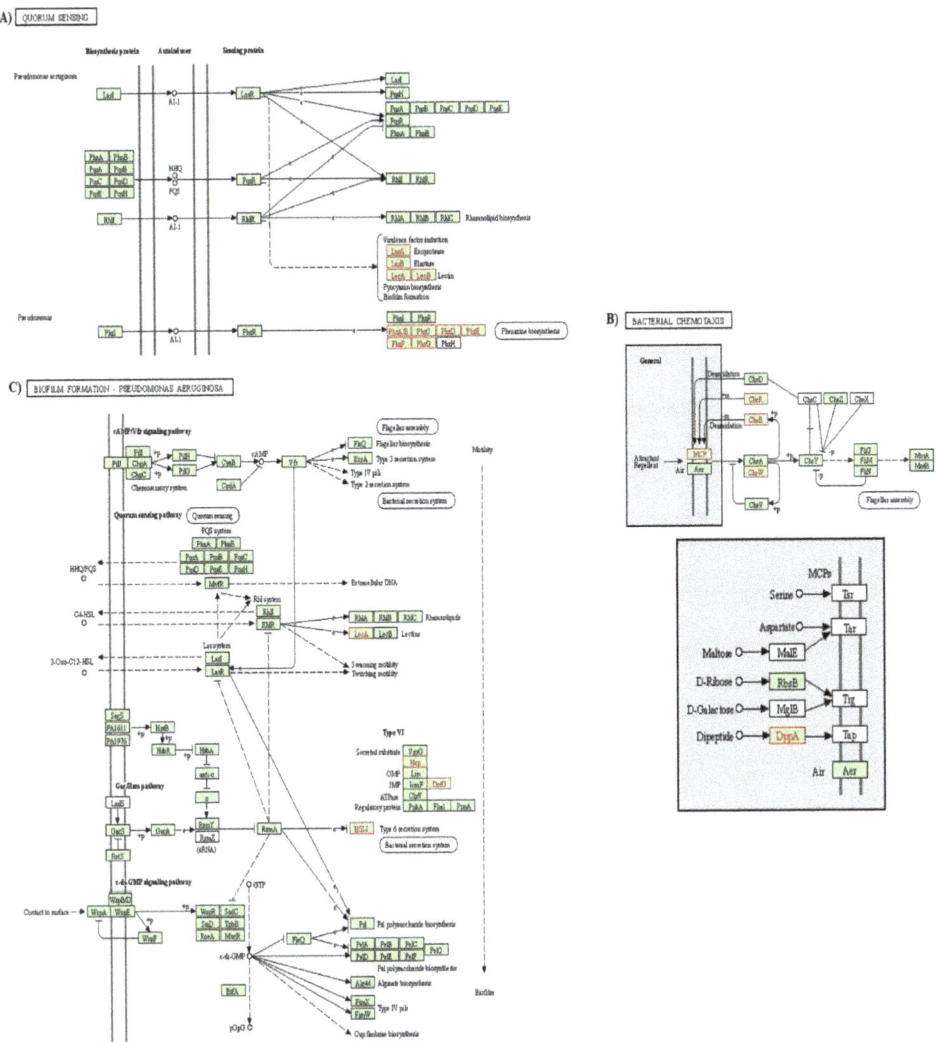

Figure 5. The KEGG pathways based on RNA high-throughput sequencing analysis of (**A**) quorum sensing, (**B**) bacterial chemotaxis, (**C**) biofilm formation. The red box indicate that the RNA expression of the gene is down-regulated, while the black box shows no changes in gene RNA expression.

Based on KEGG pathway and GO enrichment analysis, pine honey affected, at the transcriptome level, a wide range of biological processes and pathways in *P. aeruginosa*. The two-component regulatory system, the ABC transporter, and QS pathway were the most affected KEGG pathways in *P. aeruginosa*, since several up and down-regulated DEGs exhibited high fold changes (Figures 4 and 5). A two-component regulatory system plays a substantial role in the pathogenicity, bacterial adaptation, and biofilm formation [75,76]. The two-component regulatory system KEGG pathway (also called "two-component signal transduction system") enables bacteria to sense and respond to environmental or intracellular changes [77,78]. In this study, pine honey treatment induced the differential expression of several genes implicated in this pathway (Figure 4A). Among the down-regulated DEGs in the above pathway, *cheW*, *B* and *R*, PA14_02220 and *pctA* genes encode chemotaxis proteins and transducers, respectively. The *cheW, B* and *R* DEGs were also detected in the bacterial chemotaxis pathway (Figure 5B). Bacterial chemotaxis is the movement of bacterial cells in response to chemical stimuli [79]. According to Turner et al. [80], *cheW, B,* and *R* genes, are required in acute but not chronic wound infections. These data suggest that pine honey treatment might impair the two-component system and bacterial chemotaxis pathways thus reducing the ability of *P. aeruginosa* to sense environmental stimuli and adapt accordingly. In comparison to the study of Bouzo et al. [56], manuka honey treatment did not affect at the same extent the two-component regulatory system and bacterial chemotaxis pathways.

Regarding the ABC transporter pathway, pine honey treatment caused significant up-regulation of 25 genes and down-regulation of 14 genes (Figure 4B). ABC (ATP-binding cassette) transporters play an important role in nutrients uptake [81]. In addition, ABC transporter and two-component regulatory systems have a pivotal role in antimicrobial drug resistance [82]. It might be that up-regulation of several ABC transporter genes might be related to nutrient uptake directly from pine honey (e.g., sugars).

Furthermore, KEEG analysis revealed that pine honey treatment significantly inhibited QS, bacterial chemotaxis, and biofilm formation pathways, since several key genes were down-regulated (Figure 5). In *P. aeruginosa*, three systems *las, rhl,* as well as *pqs*, which are forming an hierarchical network, play a crucial role in QS [83,84]. The *las* system positively regulates itself as well as the other two systems, while the *rhl* and *pqs* systems regulate each other (Figure 5A). In the first system, *lasI* catalyzes the synthesis of the signal molecule (AI-1), by binding *lasR* and activating the expression of many genes (*pqsA, B, C, D, E, H, R, phnA, B* and *rhlI, R*) [85,86]. In the *pqs* system [87], genes such as *pqsA, B, C, D, H,* and *phnA, B,* catalyze the synthesis of the signal molecules (HHQ or PQS), by binding *pqsR* and activating the expression of various genes, including *pqsR* as well as *rhlI, R*, whereas in the third system, *rhlI* catalyzes the synthesis of the signal molecule (AI-1), by binding *rhlR* and activating the expression of other target genes (*pqsR, phnA, B, rhlI, R* and *rhlA, B*) involved in the rhamnolipid biosynthesis [88]. Furthermore, in Figure 5A it is shown that the virulence factors *lasA* (exoprotease), *lasB* (elastase) and *lecB* (lectin), pyocyanin biosynthesis, and biofilm formation are co-regulated by the three QS systems (*las, pqs,* and *rhl*). In this study, pine honey treatment inhibited the expression of virulence genes such as *lasA, lasB, pa1L* (*lecA*) and *lecB* (Figure 5A). The gene *lecA* is also involved in the biofilm formation pathway. In addition, the genes of the operon *phzABCDEFG* involved in the phenazine biosynthesis, and the genes *phzS* and *M* implicated in the pyocyanin biosynthesis, were down-regulated in a similar manner (log_2FC > 3). Previous studies have demonstrated that several enzymes of the biosynthetic operon *phzABCDEFG*, which is conserved across the fluorescent Pseudomonads, are involved in phenazine biosynthesis, through the conversion of chorismic acid to phenazine-1-carboxylic acid (PCA) [89,90]. *P. aeruginosa* has two functional copies (*phz1* and *phz2*) of this operon, which produce PCA. The conversion of PCA to phenazine-1-carboxamide as well as to 1-hydroxyphenazine is mediated by two genes *phzH* and *phzS*, respectively. A third additional gene *phzM* is involved in PCA conversion to 5-methylphenazine-1-carboxylic acid betaine, which is further converted to pyocyanin by the action of *phzS* [89–91]. A recent study showed that phenazine production is associated with the antibiotic tolerance in *P. aeruginosa* biofilms [92].

Regarding the biofilm formation pathway, KEEG analysis revealed that pine honey treatment down-regulated key genes including *pa1L* (lecA) that encode a PA-I galactophilic lectin. Additionally, several genes encoding conserved hypothetical proteins (HIS-I; PA14_34000, PA14_34030, PA14_34050, PA14_34070 and PA14_34100) were also inhibited (Figure 5C). Interestingly, pine honey treatment also down-regulated PA14_34030 (Hcp) and PA14_34110 (DotU) implicated in the type VI secretion system of *P. aeruginosa* (Figure 5C) and PA14_55940 (putative pilus assembly protein) gene of the protein secretion/export apparatus (Type II secretion system) (Figure 5C). In the bacterial chemotaxis pathway, besides *cheW, B* and *R* genes, pine honey also down-regulated PA14_58350 (DppA), PA14_61300, and PA14_02220 (MCP) (Figure 5B). Collectively, these results indicate that pine honey down-regulated several genes involved in the QS system (virulence factors, phenazine production, chemotaxis, and biofilm formation pathway) thus reducing the fitness of *P. aeruginosa* to initiate infection or biofilm formation.

In a very recent study, transcriptome analysis of *P. aeruginosa* biofilm treated with *Trigona* honey revealed that roughly 13.5% of the down-regulated genes were biofilm-associated genes. Additionally, in the pathways involved in biofilm formation, an ultimate decrease in the expression levels of the D-GMP signaling pathway and diguanylate cyclases genes implicated in c-di-GMP formation, has been observed [93].

In comparison to the study of Bouzo et al. [56], manuka honey mainly down-regulated genes of the three different QS systems (*las*, *rhl* and *pqs*) while in this study pine honey treatment demonstrated a direct inhibitory effect on genes encoding virulence factors and phenazine biosynthesis. Interestingly, pine honey also down-regulated genes implicated in bacterial chemotaxis, biofilm formation, and bacterial secretion pathways, indicating a broader mode of action on the QS system, while this does not occur at such extent following manuka honey treatment.

4. Conclusions

The present study is the first to employ a global transcriptomic approach, in order to investigate the antibacterial effects and mode of action of pine honey. RNA-seq analysis revealed that pine honey significantly affected the trascriptomic profile of *P. aeruginosa* by increasing significantly the expression of 189 genes and by reducing significantly the expression of 274 genes. Specifically, pine honey treatment exerted a broad range of action on several pathways and biological processes including oxidation-reduction process, transmembrane transport, proteolysis, regulation of DNA-templated transcription, two-component regulatory systems, ABC transporters, and SOS response. Interestingly, pine honey might inhibit quorum sensing, bacterial chemotaxis, and biofilm formation since several differentially expressed genes involved in the above pathways were strongly down-regulated. Overall, these data demonstrated that pine honey exerted an inhibitory effect in *P. aeruginosa* genome expression since more genes were down-regulated than up-regulated. These findings could potentially contribute to the treatment and control of *P. aeruginosa* infection and pathogenicity, helping to elucidate the molecular pathways and biological processes implicated in the antibacterial activity exerted by pine honey. Moreover, our results suggest that the use of pine honey in wound dressings could be an effective and economical approach to ameliorate wound healing.

Supplementary Materials: The following are available online at https://www.mdpi.com/article/10.3390/foods10050936/s1. Table S1: Differential gene expression results of HISAT2-DESeq2 package. Table S2: Differential gene expression results of de novo Trinity-edgeR package. Table S3: Differential gene expression results of reference genome guided analysis-Bowtie2-edgeR package. Table S4: Gene Ontology (all categories) of DEGs. Table S5: Gene Ontology (biological process category) of DEGs. Table S6: Gene Ontology (cellular component category) of DEGs. Table S7: Gene Ontology (molecular function category) of DEGs.

Author Contributions: I.K.: Investigation, methodology, data analysis and writing original draft. C.T.: Investigation, methodology, data analysis and writing original draft. M.N.: Data analysis and writing original draft. E.T.: Investigation and data analysis. T.G.D.: Review and editing. I.I.: Data analysis, review and editing. G.D.A.: Data analysis, review and editing. D.M.: Conceptualization, methodology, data analysis, writing original draft, review and editing, supervision, funding acquisition. All authors have read and agreed to the published version of the manuscript.

Funding: This work was supported by the General Secretariat for Research and Innovation (G.S.R.I) under the emblematic action "Honeybee Road". I.K. is recipient of a Postdoctoral Fellowship under the emblematic action "Honeybee Road". The research work was supported by the Hellenic Foundation for Research and Innovation (HFRI) under the HFRI PhD Fellowship grant (Fellowship Number: 545) for C.T.

Informed Consent Statement: Not applicable.

Data Availability Statement: The raw RNA-seq data are available at NCBI Sequence Read Archive (NCBI SRA) under BioProject accession no. PRJNA705535. https://www.ncbi.nlm.nih.gov/bioproject/?term=PRJNA705535 (accessed on 1 March 2021).

Acknowledgments: We thank Jean-Paul Pirnay (Laboratory for Molecular and Cellular Technology, Queen Astrid Military Hospital, Brussels, Belgium) who kindly provided *Pseudomonas aeruginosa* PA14 strain for the experiment.

Conflicts of Interest: The authors declare no conflict of interest.

References

1. Lee, D.J.; Jo, A.R.; Jang, M.C.; Nam, J.; Choi, H.J.; Choi, G.W.; Sung, H.Y.; Bae, H.; Ku, Y.G.; Chi, Y.T. Analysis of two quorum sensing-deficient isolates of *Pseudomonas aeruginosa*. *Microb. Pathog.* **2018**, *119*, 162–169. [CrossRef]
2. Faure, E.; Kwong, K.; Nguyen, D. *Pseudomonas aeruginosa* in chronic lung infections: How to adapt within the host? *Front. Immunol.* **2018**, *9*, 2416. [CrossRef]
3. Ruffin, M.; Brochiero, E. Repair process impairment by *Pseudomonas aeruginosa* in epithelial tissues: Major features and potential therapeutic avenues. *Front. Cell Infect. Microbiol.* **2019**, *9*, 182. [CrossRef] [PubMed]
4. El-Mahdy, R.; El-Kannishy, G. Virulence factors of carbapenem-resistant *Pseudomonas aeruginosa* in hospital-acquired infections in Mansoura, Egypt. *Infect. Drug Resist.* **2019**, *12*, 3455–3461. [CrossRef] [PubMed]
5. Nikolaidis, M.; Mossialos, D.; Oliver, S.; Amoutzias, G.D. Comparative analysis of the core proteomes among the Pseudomonas major evolutionary groups reveals species-specific adaptations for *Pseudomonas aeruginosa* and *Pseudomonas chlororaphis*. *Diversity* **2020**, *12*, 289. [CrossRef]
6. He, J.; Jia, X.; Yang, S.; Xu, X.; Sun, K.; Li, C.; Yang, T.; Zhang, L. Heteroresistance to carbapenems in invasive *Pseudomonas aeruginosa* infections. *Int. J. Antimicrob. Agents* **2018**, *51*, 413–421. [CrossRef]
7. Li, W.R.; Ma, Y.K.; Shi, Q.S.; Xie, X.B.; Sun, T.L.; Peng, H.; Huang, X.M. Diallyl disulfide from garlic oil inhibits virulence factors of *Pseudomonas aeruginosa* by inactivating key quorum sensing genes. *Appl. Microbiol. Biotechnol.* **2018**, *102*, 7555–7564. [CrossRef] [PubMed]
8. Pang, Z.; Raudonis, R.; Glick, B.R.; Lin, T.J.; Cheng, Z. Antibiotic resistance in *Pseudomonas aeruginosa*: Mechanisms and alternative therapeutic strategies. *Biotechnol. Adv.* **2019**, *37*, 177–192. [CrossRef]
9. Tacconelli, E.; Carrara, E.; Savoldi, A.; Harbarth, S.; Mendelson, M.; Monnet, D.L.; Pulcini, C.; Kahlmeter, G.; Kluytmans, J.; Carmeli, Y.; et al. Discovery, research, and development of new antibiotics: The WHO priority list of antibiotic-resistant bacteria and tuberculosis. *Lancet Infect. Dis.* **2018**, *18*, 318–327. [CrossRef]
10. McLoone, P.; Warnock, M.; Fyfe, L. Honey: A realistic antimicrobial for disorders of the skin. *J. Microbiol. Immunol. Infect.* **2016**, *49*, 161–167. [CrossRef] [PubMed]
11. Oryan, A.; Alemzadeh, E.; Moshiri, A. Biological properties and therapeutic activities of honey in wound healing: A narrative review and meta-analysis. *J. Tissue Viability* **2016**, *25*, 98–118. [CrossRef]
12. Khan, S.U.; Anjum, S.I.; Rahman, K.; Ansari, M.J.; Khan, W.U.; Kamal, S.; Khattak, B.; Muhammad, A.; Khan, H.U. Honey: Single food stuff comprises many drugs. *Saudi J. Biol. Sci.* **2018**, *25*, 320–325. [CrossRef]
13. Roshan, N.; Rippers, T.; Locher, C.; Hammer, K.A. Antibacterial activity and chemical characteristics of several Western Australian honeys compared to manuka honey and pasture honey. *Arch. Microbiol.* **2017**, *199*, 347–355. [CrossRef] [PubMed]
14. Kuś, P.M.; Szweda, P.; Jerković, I.; Tuberoso, C.I. Activity of Polish unifloral honeys against pathogenic bacteria and its correlation with colour, phenolic content, antioxidant capacity and other parameters. *Lett. Appl. Microbiol.* **2016**, *62*, 269–276. [CrossRef] [PubMed]
15. Grecka, K.; Kuś, P.M.; Worobo, R.W.; Szweda, P. Study of the anti-staphylococcal potential of honeys produced in Northern Poland. *Molecules* **2018**, *23*, 260. [CrossRef] [PubMed]
16. Bucekova, M.; Jardekova, L.; Juricova, V.; Bugarova, V.; Di Marco, G.; Gismondi, A.; Leonardi, D.; Farkasovska, J.; Godocikova, J.; Laho, M.; et al. Antibacterial activity of different blossom honeys: New findings. *Molecules* **2019**, *24*, 1573. [CrossRef]
17. Bucekova, M.; Bugárová, V.; Godocikova, J.; Majtan, J. Demanding new honey qualitative standard based on antibacterial activity. *Foods* **2020**, *9*, 1263. [CrossRef] [PubMed]
18. Anthimidou, E.; Mossialos, D. Antibacterial activity of Greek and Cypriot honeys against *Staphylococcus aureus* and *Pseudomonas aeruginosa* in comparison to manuka honey. *J. Med. Food* **2013**, *16*, 42–47. [CrossRef]
19. Stagos, D.; Soulitsiotis, N.; Tsadila, C.; Papaeconomou, S.; Arvanitis, C.; Ntontos, A.; Karkanta, F.; Androulaki, S.A.; Petrotos, K.; Spandidos, D.A.; et al. Antibacterial and antioxidant activity of different types of honey derived from Mount Olympus in Greece. *Int. J. Mol. Med.* **2018**, *42*, 726–734. [CrossRef]
20. Tsavea, E.; Mossialos, D. Antibacterial activity of honeys produced in Mount Olympus area against nosocomial and foodborne pathogens is mainly attributed to hydrogen peroxide and proteinaceous compounds. *J. Apic. Res.* **2019**, *58*, 1–8. [CrossRef]
21. Kwakman, P.H.S.; Zaat, S. Antibacterial components of honey. *IUBMB Life* **2012**, *64*, 48–55. [CrossRef] [PubMed]
22. Nolan, V.C.; Harrison, J.; Cox, J. Dissecting the antimicrobial composition of honey. *Antibiotics* **2019**, *8*, 251. [CrossRef] [PubMed]
23. Jenkins, R.; Burton, N.; Cooper, R. Effect of manuka honey on the expression of universal stress protein A in methicillin- resistant *Staphylococcus aureus*. *Int. J. Antimicrob. Agents* **2011**, *37*, 373–376. [CrossRef] [PubMed]

24. Sindi, A.; Chawn, M.V.B.; Hernandez, M.E.; Green, K.; Islam, M.K.; Locher, C.; Hammer, K.A. Anti-biofilm effects and characterisation of the hydrogen peroxide activity of a range of Western Australian honeys compared to manuka and multifloral honeys. *Sci. Rep.* **2019**, *9*, 17666. [CrossRef]

25. Roberts, A.E.; Maddocks, S.E.; Cooper, R.A. Manuka honey reduces the motility of *Pseudomonas aeruginosa* by suppression of flagella-associated genes. *J. Antimicrob. Chemother.* **2015**, *70*, 716–725. [CrossRef]

26. Ahmed, A.A.; Salih, F.A. Low concentrations of local honey modulate Exotoxin A expression, and quorum sensing related virulence in drug-resistant *Pseudomonas aeruginosa* recovered from infected burn wounds. *Iran J. Basic Med. Sci.* **2019**, *22*, 568–575. [CrossRef] [PubMed]

27. Nikolopoulos, C. *Morphology and Biology of the Species Marchalina Hellenica (Gennadius) (Hemiptera, Margarodidea-Coelostomidiinae)*; Agricultural College of Athens: Athens, Greece, 1965; p. 16.

28. de-Miguel, S.; Pukkala, T.; Yeşil, A. Integrating pine honeydew honey production into forest management optimization. *Eur. J. For. Res.* **2014**, *133*, 423–432. [CrossRef]

29. Tananaki, C.H.; Thrasyvoulou, A.; Giraudel, J.L.; Montury, M. Determination of volatile characteristics of Greek and Turkish pine honey samples and their classification by using Kohonen selforganizing maps. *Food Chem.* **2007**, *101*, 1687–1693. [CrossRef]

30. Bacandritsos, N.; Saitanis, C.; Papanastasiou, I. Morphology and life cycle of *Marchalina hellenica* (Gennadius) (Hemiptera: Margarodidae) on pine (Parnis Mt.) and fir (Helmos Mt.) forests of Greece. *Ann. Soc. Entomol. Fr.* **2004**, *40*, 169–176. [CrossRef]

31. Bouga, M.; Evangelou, V.; Lykoudis, D.; Cakmak, I.; Hatjina, F. Genetic structure of *Marchalina hellenica* (Hemiptera: Margarodidae) populations from Turkey: Preliminary mtDNA sequencing data. *Biochem. Genet.* **2011**, *49*, 683–694. [CrossRef]

32. Akbulut, M.; Ozcan, M.M.; Coklar, H. Evaluation of antioxidant activity, phenolic, mineral contents and some physicochemical properties of several pine honeys collected from Western Anatolia. *Int. J. Food Sci. Nutr.* **2009**, *60*, 577–589. [CrossRef]

33. Kaygusuz, H.; Tezcan, F.; Bedia Erim, F.; Yildiz, O.; Sahin, H.; Can, Z.; Kolayli, S. Characterization of Anatolian honeys based on minerals, bioactive components and principal component analysis. *LWT-Food Sci. Technol.* **2016**, *68*, 273–279. [CrossRef]

34. Kolayli, S.; Sahin, H.; Can, Z.; Yildiz, O.; Sahin, K. Honey shows potent inhibitory activity against the bovine testes hyaluronidase. *Enzyme Inhib. Med. Chem.* **2016**, *31*, 599–602. [CrossRef] [PubMed]

35. Gül, A.; Pehlivan, T. Antioxidant activities of some monofloral honey types produced across Turkey. *Saudi J. Biol. Sci.* **2018**, *25*. [CrossRef]

36. Alnaimat, S.; Wainwright, M.; Al'Abri, K. Antibacterial potential of honey from different origins: A comparsion with manuka honey. *J. Microbiol. Biotechnol. Food Sci.* **2012**, *1*, 1328–1338.

37. Ekici, L.; Osman, S.; Sibel, S.; Ismet, O. Determination of phenolic content, antiradical, antioxidant and antimicrobial activities of Turkish pine honey. *Qual. Assur. Saf. Crop. Foods* **2014**, *6*, 439–444. [CrossRef]

38. Lei, R.; Ye, K.; Gu, Z.; Sun, X. Diminishing returns in next-generation sequencing (NGS) transcriptome data. *Gene* **2015**, *557*, 82–87. [CrossRef] [PubMed]

39. Łabaj, P.P.; Kreil, D.P. Sensitivity, specificity, and reproducibility of RNA-Seq differential expression calls. *Biol. Direct.* **2016**, *11*, 66. [CrossRef]

40. Liu, X.; Shen, B.; Du, P.; Wang, N.; Wang, J.; Li, J.; Sun, A. Transcriptomic analysis of the response of *Pseudomonas fluorescens* to epigallocatechin gallate by RNA-seq. *PLoS ONE* **2017**, *12*, e0177938. [CrossRef]

41. Li, Z.; Xu, M.; Wei, H.; Wang, L.; Deng, M. RNA-seq analyses of antibiotic resistance mechanisms in *Serratia marcescens*. *Mol. Med. Rep.* **2019**, *20*, 745–754. [CrossRef] [PubMed]

42. Szweda, P. Antimicrobial activity of honey. In *Honey Analysis*; Arnaut de Toledo, V.A., Ed.; InTech: Madrid, Spain, 2017; pp. 215–232. [CrossRef]

43. Kwakman, P.H.S.; Te Velde, A.A.; de Boer, L.; Speijer, D.; Vandenbroucke-Grauls, C.M.; Zaat, S.A.J. How honey kills bacteria. *FASEB J.* **2010**, *24*, 2576–2582. [CrossRef]

44. Bolger, A.M.; Lohse, M.; Usadel, B. Trimmomatic: A flexible trimmer for Illumina sequence data. *Bioinformatics* **2014**, *30*, 2114–2120. [CrossRef]

45. Kim, D.; Langmead, B.; Salzberg, S.L. HISAT: A fast spliced aligner with low memory requirements. *Nat. Methods* **2015**, *12*, 357–360. [CrossRef]

46. Liao, Y.; Smyth, G.K.; Shi, W. FeatureCounts: An efficient general purpose program for assigning sequence reads to genomic features. *Bioinformatics* **2013**, *30*, 923–930. [CrossRef]

47. Love, M.I.; Huber, W.; Anders, S. Moderated estimation of fold change and dispersion for RNA-seq data with DESeq2. *Genome Biol.* **2014**, *15*, 550. [CrossRef]

48. Young, M.D.; Wakefield, M.J.; Smyth, G.K.; Oshlack, A. Gene ontology analysis for RNA-seq: Accounting for selection bias. *Genome Biol.* **2010**, *11*, R14. [CrossRef] [PubMed]

49. Kanehisa, M.; Goto, S.; Furumichi, M.; Tanabe, M.; Hirakawa, M. KEGG for representation and analysis of molecular networks involving diseases and drugs. *Nucleic Acids Res.* **2010**, *38*, D355–D360. [CrossRef] [PubMed]

50. Winsor, G.L.; Griffiths, E.J.; Lo, R.; Dhillon, B.K.; Shay, J.A.; Brinkman, F. Enhanced annotations and features for comparing thousands of Pseudomonas genomes in the Pseudomonas genome database. *Nucleic Acids Res.* **2016**, *44*, D646–D653. [CrossRef] [PubMed]

51. Grabherr, M.G.; Haas, B.J.; Yassour, M.; Levin, J.Z.; Thompson, D.A.; Amit, I.; Adiconis, X.; Fan, L.; Raychowdhury, R.; Zeng, Q.; et al. Full-length transcriptome assembly from RNA-Seq data without a reference genome. *Nat. Biotechnol.* **2011**, *29*, 644–652. [CrossRef] [PubMed]

52. Langmead, B.; Salzberg, S.L. Fast gapped-read alignment with Bowtie 2. *Nat. Methods* **2012**, *9*, 357–359. [CrossRef]

53. Robinson, M.D.; McCarthy, D.J.; Smyth, G.K. edgeR: A bioconductor package for differential expression analysis of digital gene expression data. *Bioinformatics* **2010**, *26*, 139–140. [CrossRef]

54. Wecke, T.; Mascher, T. Antibiotic research in the age of omics: From expression profiles to interspecies communication. *J. Antimicrob. Chemother.* **2011**, *66*, 2689–2704. [CrossRef] [PubMed]

55. Bernier, S.P.; Surette, M.G. Concentration-dependent activity of antibiotics in natural environments. *Front. Microbiol.* **2013**, *4*, 20. [CrossRef] [PubMed]

56. Bouzo, D.; Cokcetin, N.; Li, L.; Ballerin, G.; Bottomley, A.; Lazenby, J.; Whitchurch, C.; Paulsen, I.; Hassan, K.; Harry, E. Characterizing the mechanism of action of an ancient antimicrobial, manuka honey, against *Pseudomonas aeruginosa* using modern transcriptomics. *MSystems* **2020**, *5*, e00106-20. [CrossRef]

57. Brown, S.M.; Howell, M.L.; Vasil, M.L.; Anderson, A.J.; Hassett, D.J. Cloning and characterization of the katB gene of *Pseudomonas aeruginosa* encoding a hydrogen peroxide-inducible catalase: Purification of katB, cellular localization, and demonstration that it is essential for optimal resistance to hydrogen peroxide. *J. Bacteriol.* **1995**, *177*, 6536–6544. [CrossRef] [PubMed]

58. Mossialos, D.; Tavankar, G.; Zlosnik, J.; Williams, H. Defects in a quinol oxidase lead to loss of katC catalase activity in *Pseudomonas aeruginosa*: KatC activity is temperature dependent and it requires an intact disulphide bond formation system. *Biochem. Biophys. Res. Commun.* **2006**, *341*, 697–702. [CrossRef] [PubMed]

59. Wongsaroj, L.; Saninjuk, K.; Romsang, A.; Duang-Nkern, J.; Trinachartvanit, W.; Vattanaviboon, P.; Mongkolsuk, S. *Pseudomonas aeruginosa* glutathione biosynthesis genes play multiple roles in stress protection, bacterial virulence and biofilm formation. *PLoS ONE* **2018**, *13*, e0205815. [CrossRef] [PubMed]

60. Tang, D.; Wang, X.; Wang, J.; Wang, M.; Wang, Y.; Wang, W. Choline-betaine pathway contributes to hyperosmotic stress and subsequent lethal stress resistance in *Pseudomonas protegens* SN15-2. *J. Biosci.* **2020**, *45*, 85. [CrossRef]

61. Ashburner, M.; Ball, C.A.; Blake, J.A.; Botstein, D.; Butler, H.; Cherry, J.M.; Davis, A.P.; Dolinski, K.; Dwight, S.S.; Eppig, J.T.; et al. Gene ontology: Tool for the unification of biology. The Gene Ontology Consortium. *Nat. Genet.* **2000**, *25*, 25–29. [CrossRef]

62. Gao, P.; Guo, K.; Pu, Q.; Wang, Z.; Lin, P.; Qin, S.; Khan, N.; Hur, J.; Liang, H.; Wu, M. OprC impairs host defense by increasing the quorum-sensing-mediated virulence of *Pseudomonas aeruginosa*. *Front. Immunol.* **2020**, *11*, 1696. [CrossRef] [PubMed]

63. Wang, R.; Starkey, M.; Hazan, R.; Rahme, L.G. Honey's ability to counter bacterial infections arises from both bactericidal compounds and QS inhibition. *Front. Microbiol.* **2012**, *3*, 144. [CrossRef] [PubMed]

64. Pirnay, J.P.; De Vos, D.; Mossialos, D.; Vanderkelen, A.; Cornelis, P.; Zizi, M. Analysis of the *Pseudomonas aeruginosa* oprD gene from clinical and environmental isolates. *Environ. Microbiol.* **2002**, *4*, 872–882. [CrossRef]

65. Hancock, R.E.; Speert, D.P. Antibiotic resistance in *Pseudomonas aeruginosa*: Mechanisms and impact on treatment. *Drug Resist. Updat.* **2000**, *3*, 247–255. [CrossRef]

66. Adewoye, L.O.; Worobec, E.A. Identification and characterization of the gltK gene encoding a membrane-associated glucose transport protein of *Pseudomonas aeruginosa*. *Gene* **2000**, *253*, 323–330. [CrossRef]

67. Raneri, M.; Pinatel, E.; Peano, C.; Rampioni, G.; Leoni, L.; Bianconi, I.; Jousson, O.; Dalmasio, C.; Ferrante, P.; Briani, F. *Pseudomonas aeruginosa* mutants defective in glucose uptake have pleiotropic phenotype and altered virulence in non-mammal infection models. *Sci. Rep.* **2018**, *8*, 16912. [CrossRef]

68. Lister, P.D.; Wolter, D.J.; Hanson, N.D. Antibacterial-resistant *Pseudomonas aeruginosa*: Clinical impact and complex regulation of chromosomally encoded resistance mechanisms. *Clin. Microbiol. Rev.* **2009**, *22*, 582–610. [CrossRef] [PubMed]

69. Quale, J.; Bratu, S.; Gupta, J.; Landman, D. Interplay of efflux system, ampC, and oprD expression in carbapenem resistance of *Pseudomonas aeruginosa* clinical isolates. *Antimicrob. Agents Chemother.* **2006**, *50*, 1633–1641. [CrossRef] [PubMed]

70. Mossialos, D.; Amoutzias, G.D. Siderophores in fluorescent pseudomonads: New tricks from an old dog. *Future Microbiol.* **2007**, *2*, 387–395. [CrossRef]

71. Budzikiewicz, H. Siderophore-antibiotic conjugates used as trojan horses against *Pseudomonas aeruginosa*. *Curr. Top. Med. Chem.* **2001**, *1*, 73–82. [CrossRef] [PubMed]

72. Mossialos, D.; Amoutzias, G.D. Role of siderophores in cystic fibrosis pathogenesis: Foes or friends? *Int. J. Med. Microbiol.* **2009**, *299*, 87–98. [CrossRef]

73. Mislin, G.L.; Schalk, I.J. Siderophore-dependent iron uptake systems as gates for antibiotic Trojan horse strategies against *Pseudomonas aeruginosa*. *Metallomics* **2014**, *6*, 408–420. [CrossRef]

74. Luscher, A.; Moynié, L.; Auguste, P.S.; Bumann, D.; Mazza, L.; Pletzer, D.; Naismith, J.H.; Köhler, T. TonB-dependent receptor repertoire of *Pseudomonas aeruginosa* for uptake of siderophore-drug conjugates. *Antimicrob. Agents Chemother.* **2018**, *62*, e00097-18. [CrossRef]

75. Francis, V.I.; Stevenson, E.C.; Porter, S.L. Two-component systems required for virulence in *Pseudomonas aeruginosa*. *FEMS Microbiol. Lett.* **2017**, *364*, fnx104. [CrossRef] [PubMed]

76. Bhagirath, A.Y.; Li, Y.; Patidar, R.; Yerex, K.; Ma, X.; Kumar, A.; Duan, K. Two component regulatory systems and antibiotic resistance in Gram-negative pathogens. *Int. J. Mol. Sci.* **2019**, *20*, 1781. [CrossRef]

77. Tiwari, S.; Jamal, S.B.; Hassan, S.S.; Carvalho, P.; Almeida, S.; Barh, D.; Ghosh, P.; Silva, A.; Castro, T.; Azevedo, V. Two-component signal transduction systems of pathogenic bacteria as targets for antimicrobial therapy: An overview. *Front. Microbiol.* **2017**, *8*, 1878. [CrossRef] [PubMed]

78. Hirakawa, H.; Kurushima, J.; Hashimoto, Y.; Tomita, H. Progress overview of bacterial two-component regulatory systems as potential targets for antimicrobial chemotherapy. *Antibiotics* **2020**, *9*, 635. [CrossRef]

79. Hazelbauer, G.L. Bacterial chemotaxis: The early years of molecular studies. *Annu. Rev. Microbiol.* **2012**, *66*, 285–303. [CrossRef]

80. Turner, K.H.; Everett, J.; Trivedi, U.; Rumbaugh, K.P.; Whiteley, M. Requirements for *Pseudomonas aeruginosa* acute burn and chronic surgical wound infection. *PLoS Genet.* **2014**, *10*, e1004518. [CrossRef]

81. Pletzer, D.; Braun, Y.; Dubiley, S.; Lafon, C.; Köhler, T.; Page, M.; Mourez, M.; Severinov, K.; Weingart, H. The *Pseudomonas aeruginosa* PA14 ABC transporter NppA1A2BCD is required for uptake of peptidyl nucleoside antibiotics. *J. Bacteriol.* **2015**, *197*, 2217–2228. [CrossRef] [PubMed]

82. Ahmad, A.; Majaz, S.; Nouroz, F. Two-component systems regulate ABC transporters in antimicrobial peptide production, immunity and resistance. *Microbiology* **2020**, *166*, 4–20. [CrossRef] [PubMed]

83. Guo, Q.; Kong, W.N.; Jin, S.; Chen, L.; Xu, Y.Y.; Duan, K.M. PqsR-dependent and PqsR-independent regulation of motility and biofilm formation by PQS in *Pseudomonas aeruginosa* PAO1. *J. Basic Microbiol.* **2014**, *54*, 633–643. [CrossRef] [PubMed]

84. Lee, J.; Zhang, L. The hierarchy quorum sensing network in *Pseudomonas aeruginosa*. *Protein Cell* **2015**, *6*, 26–41. [CrossRef] [PubMed]

85. Passador, L.; Cook, J.M.; Gambello, M.J.; Rust, L.; Lglewski, B.H. Expression of *Pseudomonas aeruginosa* virulence genes requires cell-to-cell communication. *Science* **1993**, *260*, 1127–1130. [CrossRef]

86. Finch, R.G.; Pritchard, D.I.; Bycroft, B.W.; Williams, P.; Stewart, G.S. Quorum sensing: A novel target for anti-infective therapy. *J. Antimicrob. Chemother.* **1998**, *42*, 569–571. [CrossRef]

87. Williams, P.; Cámara, M. Quorum sensing and environmental adaptation in *Pseudomonas aeruginosa*: A tale of regulatory networks and multifunctional signal molecules. *Curr. Opin. Microbiol.* **2009**, *12*, 182–191. [CrossRef] [PubMed]

88. Latifi, A.; Foglino, M.; Tanaka, K.; Williams, P.; Lazdunski, A. A hierarchical quorum-sensing cascade in *Pseudomonas aeruginosa* links the transcriptional activators LasR and RhlR (VsmR) to expression of the stationary-phase sigma factor RpoS. *Mol. Microbiol.* **1996**, *21*, 1137–1146. [CrossRef]

89. Mavrodi, D.V.; Blankenfeldt, W.; Thomashow, L.S. Phenazine compounds in fluorescent *Pseudomonas* spp. biosynthesis and regulation. *Ann. Rev. Phytopathol.* **2006**, *44*, 417–445. [CrossRef]

90. Mavrodi, D.V.; Peever, T.L.; Mavrodi, O.V.; Parejko, J.A.; Raaijmakers, J.M.; Lemanceau, P.; Mazurier, S.; Heide, L.; Blankenfeldt, W.; Weller, D.M.; et al. Diversity and evolution of the phenazine biosynthesis pathway. *Appl. Environ. Microbiol.* **2010**, *76*, 866–879. [CrossRef]

91. Mavrodi, D.V.; Bonsall, R.F.; Delaney, S.M.; Soule, M.J.; Phillips, G.; Thomashow, L.S. Functional analysis of genes for biosynthesis of pyocyanin and phenazine-1-carboxamide from *Pseudomonas aeruginosa* PAO1. *J. Bacteriol.* **2001**, *183*, 6454–6465. [CrossRef]

92. Schiessl, K.T.; Hu, F.; Jo, J.; Nazia, S.; Wang, B.; Price-Whelan, A.; Min, W.; Dietrich, L. Phenazine production promotes antibiotic tolerance and metabolic heterogeneity in *Pseudomonas aeruginosa* biofilms. *Nat. Commun.* **2019**, *10*, 762. [CrossRef]

93. Seder, N.; Abu Bakar, M.H.; Abu Rayyan, W.S. Transcriptome analysis of *Pseudomonas aeruginosa* biofilm following the exposure to Malaysian stingless bee honey. *Adv. Appl. Bioinform. Chem.* **2021**, *14*, 1–11. [CrossRef] [PubMed]

Article

Utilizing Pork Exudate Metabolomics to Reveal the Impact of Aging on Meat Quality

Qianqian Yu [1,2], Bruce Cooper [3], Tiago Sobreira [3] and Yuan H. Brad Kim [1,*]

1   Meat Science and Muscle Biology Laboratory, Department of Animal Science, Purdue University, West Lafayette, IN 47906, USA; yuqianqianlly@hotmail.com
2   College of Life Science, Yantai University, No. 30 Qingquan Road, Laishan District, Yantai 264005, China
3   Bindley Bioscience Center, Purdue University, West Lafayette, IN 47907, USA; brcooper@purdue.edu (B.C.); sobreira@purdue.edu (T.S.)
*   Correspondence: bradkim@purdue.edu; Tel.: +1-765-496-1631

Abstract: This study was performed to assess the changes in meat quality and metabolome profiles of meat exudate during postmortem aging. At 24 h postmortem, longissimus lumborum muscles were collected from 10 pork carcasses, cut into three sections, and randomly assigned to three aging period groups (2, 9, and 16 d). Meat quality and chemical analyses, along with the metabolomics of meat exudates using ultra-high-performance liquid chromatography coupled with a quadrupole time-of-flight mass spectrometer (UHPLC-QTOF-MS) platform, were conducted. Results indicated a declined ($p < 0.05$) display color stability, and increased ($p < 0.05$) purge loss, meat tenderness, and lipid oxidation as aging extended. The principal component analysis and hierarchical clustering analysis exhibited distinct clusters of the exudate metabolome of each aging treatment. A total of 39 significantly changed features were tentatively identified via matching them to METLIN database according to their MS/MS information. Some of those features are associated with adenosine triphosphate metabolism (creatine and hypoxanthine), antioxidation (oxidized glutathione and carnosine), and proteolysis (dipeptides and tripeptides). The findings provide valuable information that reflects the meat quality's attributes and could be used as a source of potential biomarkers for predicting aging times and meat quality changes.

Keywords: pork purge; aging; metabolome; quadrupole time-of-flight mass spectrometer (UHPLC-QTOF-MS)

Citation: Yu, Q.; Cooper, B.; Sobreira, T.; Kim, Y.H.B. Utilizing Pork Exudate Metabolomics to Reveal the Impact of Aging on Meat Quality. *Foods* **2021**, *10*, 668. https://doi.org/10.3390/foods10030668

Academic Editors: Yelko Rodríguez-Carrasco and Bojan Šarkanj

Received: 27 February 2021
Accepted: 16 March 2021
Published: 20 March 2021

Publisher's Note: MDPI stays neutral with regard to jurisdictional claims in published maps and institutional affiliations.

1. Introduction

Postmortem aging is a value-adding process and has been extensively practiced by the global meat industry for years to enhance meat tenderness, juiciness, and flavor development [1]. In particular, wet-aging (storing fresh meat at refrigerated temperatures under vacuum packaged bags) is a widely used strategy in the meat industry due to its enhanced ease and flexibility of storage [2]. However, an increase in exudate from meat during aging in a vacuum bag is unavoidable [3,4]. The main cause for the exudate release is a series of breakdowns in the myofibrils, the cell membrane structure, and the intracellular cytoskeleton during aging process [5]. In addition, physical pressure applied during vacuum packing causes the extraction of liquid from the meat, resulting in a greater release of exudation and purge collection in the bag [6]. The purge contains numerous metabolites, which could provide useful background biochemical information and could potentially be closely associated with meat quality changes. Purge, however, is mostly considered to be waste, and the potential of meat purge use for meat quality determination has not been actively realized. Only a few studies have explored the potential value of meat purge as an analytical medium.

Meat purge is the aqueous solution that contains water-soluble sarcoplasmic proteins, nucleotides, amino acids, peptides, and many soluble enzymes, which are all necessary for the metabolism of muscle and for postmortem changes in meat [7,8]. Research regarding purge loss has mainly focused on identifying exudation amount or correlating its relationship with the meat's capacity to hold water, meat tenderness, or protein oxidation [9–11]. Very limited research has been done regarding the complete profiling of the compounds in meat exudate itself. Recently, the metabolomes in beef [7] and pork [12] exudates during different aging periods were first characterized by ^{1}H nuclear magnetic resonance (NMR) spectroscopy, and the results indicated a strong correlation between exudate and meat spectra. Moreover, the metabolite profiles of the muscle exudate obtained from chicken breast fillets affected by wooden breast (WB) myopathy were revealed by NMR-based metabolomics (eleven metabolites were significantly affected by WB myopathy) [13]. These researches display valuable information about the metabolome profile changes of meat exudate and suggest an easy method to monitor metabolic profile changes associated with meat aging or meat quality by determining exudate spectra.

Mass spectrometry (MS) has emerged as the critical technology in metabolomics studies due to its unparalleled sensitivity and specificity, high resolution and wide dynamic range, enabling the comprehensive quantitative and qualitative measurement of large-scale small-molecular metabolites in complex biological samples [14]. It has been widely applied to characterize and quantify the component changes in various muscles [15–18] during postmortem aging. However, most published studies only focus on muscle tissue rather than its purge solution. In our previous study, metabolites present in exudates from beef loins and tenderloins at different aging periods (9, 16, and 23 d) were characterized using ultra-high-performance liquid chromatography coupled with electrospray ionization mass spectrometer (UPLC-ESI-MS), and the relationship between identified metabolites and the oxidative stability of beef muscles during aging was examined as well [19]. The results indicated that both the oxidative stability of meat and well as the metabolomics profile of meat exudate were significantly affected by muscle type and aging period. Furthermore, some oxidative stability-related compounds in meat exudate were identified, which could possibly explain changes in meat quality.

Taken together, the objective of the study was to assess the meat quality attribute changes and the associated purge metabolome profiles during vacuum (wet) aging in order to characterize the major metabolites that present in pork exudate with different aging periods for monitoring aging times and potentially determining changes in meat quality.

2. Materials and Methods

2.1. Sample Collection

A total of 10 pork carcasses (Landrace × Large White breed with an average live weight of 117.3 ± 1.7 kg) were harvested at the Purdue University Meat Laboratory. Longissimus lumborum muscles from one side of each carcass were collected at 24 h postmortem, then split in three sections, vacuum packaged, and randomly assigned to three aging times (2, 9, and 16 days) at 2 °C. Upon the completion of each aging period, the vacuum packages were opened and the exudates were collected and stored at −80 °C until used for metabolomics analysis. The weight of each section before and after aging were recorded in order to calculate purge loss. Pork sections were divided into multiple cuts, and two cuts (around 2.54 cm thick) of each section were used for shear force and display color evaluation, respectively. The rest of the cuts were vacuum packaged individually and stored at −80 °C until chemical analyses were conducted.

2.2. pH Measurement

The pH of the pork samples was measured at 24 h postmortem and after each assigned aging time using a meat pH probe (HANNA HI 99163, Hanna Instrument, Inc., Warner, NH, USA). The pH probe was calibrated to a pH 4 and 7 buffer at 2 °C before use and was directly inserted into three random locations of each sample for measurement.

2.3. Water-Holding Capacity (WHC)

Multiple measurements (including purge loss, drip loss, and cook loss) were conducted to determine the weight differences under various physical stressors to evaluate the WHC of the pork samples. Purge loss was measured and calculated in accordance to Setyabrata et al. [20]. Drip loss was performed according to the method obtained from Kim et al. [21]. To evaluate cooking loss, the pork cuts were first weighed and then cooked at 135 °C on a flat top electric griddle (Farberware, Walter Idde and Co., Bronx, NY, USA). Each cut was turned to its other side when the core temperature reached 41 °C and then cooked to 71 °C. After cooking, the sample's weight was recorded and the cooking loss was calculated according the following formula:

$$\text{Cooking loss (\%)} = 100 \times (1 - \frac{weight\ of\ cooked\ sample}{weight\ of\ raw\ sample})$$

2.4. Display Color

One cut from each assigned aging section was used for display color assessment. Briefly, pork cuts were aerobically packaged with polyvinyl chloride film (23,000 $cm^3/m^2/24$ h oxygen transmission rate at 23 °C) and displayed for 5 days at 2 °C under fluorescent light ($\sim$1450 lx; color temperature = 3500 K). Color was recorded daily in five random locations on the pork cuts using a Minolta CR-400 colorimeter (Konica Minolta Photo Imaging Inc., Tokyo, Japan) with CIE standard illuminant D65. The chroma and hue angle were calculated according to the following formulae described by the American Meat Science Association guidelines [22]:

$$chroma = \sqrt{(a^{*2} + b^{*2})}$$
$$\text{hue angle} = \arctan(b^*/a^*)$$

2.5. Shear Force

The cooked pork cuts were used for shear force evaluation after 24 h of cooling at 4 °C. Six cores parallel to the fiber direction were collected using a 1.27 cm diameter hand-held coring device. The measurement was performed utilizing a TA-XT Plus Texture Analyzer (Stable Micro System Ltd., Godalming, UK) coupled with a Warner–Bratzler shear attachment. The average peak shear force (Kg) from the cores was calculated for analysis.

2.6. Myofibril Fragmentation Index (MFI)

MFI was performed in accordance with Culler et al. [23], with some modifications. Two grams of the minced sample were homogenized with 20 mL cold (2 °C) MFI buffer (100 mM potassium chloride, 20 mM potassium phosphate, 1 mM ethylene glycol tetraacetic acid, 1 mM magnesium chloride, and 1 mM sodium azide; pH 7.0) for 45 s. The mixture was centrifuged at 1000× g for 15 min (at 4 °C) and then the supernatant was removed. The pellet was resuspended in 20 mL MFI buffer by using a stir rod and centrifuged again at 1000× g for 15 min. The supernatant was removed and a pellet of it was resuspended using 5 mL of cold (2 °C) MFI buffer. Straining sample using a No. 18 polyethylene strainer and an additional 5 mL buffer were used to facilitate the passage of myofibrils through the strainer. The protein concentration of the suspension was diluted with an MFI buffer to a final volume of 0.5 mg/mL and the absorbance was then measured using a UV spectrophotometer (VWR UV-1600 PC, VWR International, San Francisco, CA, USA) at 540 nm. The MFI was expressed as the absorbance multiplied by 200.

2.7. Lipid Oxidation

Lipid oxidation was evaluated via 2-thiobarbituric reactive substances (TBARS) assay according to the method outlined by Setyabrata and Kim [20]. Briefly, 5 g meat samples were homogenized with 15 mL of deionized distilled water and 50 µL of 10% butylated hydroxyl anisole solution in 90% ethanol for 30 s at 6000 rpm. One milliliter of the homogenate was mixed with 2 mL of 20 mM 2-thiobarbituric acid solution in 15% trichloroacetic acid, heated in 80 °C for 15 min, and cooled in ice for 10 min. Afterwards, the mixture was centrifuged at 2000× g for 10 min and filtered. The absorbance of the filtrate was read at 531 nm using a microplate spectrophotometer (BioTek Instruments, Inc., Winooski, VT, USA) and multiplied by 5.54 to express the TBARS value.

2.8. Protein Carbonyl Content

The carbonyl content of the pork samples was obtained through following a procedure described by Vossen and De Smet [24]. The carbonyl content was determined via derivatization with 2,4-dinitrophenylhydrazine (DNPH), leading to the formation of stable dinitrophenyl (DNP) hydrazone adducts, which can be detected spectrophotometrically at 375 nm using the Epoch Spectrophotometer System (BioTek Instruments, Inc., Winooski, VT, USA). The carbonyl concentration was expressed as nmol/mg protein.

2.9. Metabolomics Analysis for Pork Purge

2.9.1. Metabolite Extraction

Purge metabolite extraction was performed according to the method from Bligh and Dyer [25]. Chloroform (250 µL) and methanol (225 µL) were added to 100 µL of meat purge, and then the samples were vortexed for 15 s and left to sit for 5 min. Afterwards, 125 µL of water was added, mixed, and centrifuged at 16,000× g for 8 min. The upper layer was transferred to vials and dried via a termovap sample concentrator. For ultra-high-performance liquid chromatography coupled with a quadrupole time-of-flight mass spectrometer (UHPLC-QTOF-MS) analysis, the dried fraction was redissolved in 100 µL of acetonitrile aqueous solution (water:acetonitrile = 95:5) containing 0.1% formic acid.

2.9.2. UHPLC-QTOF-MS Analysis

An Agilent 1290 Infinity II UHPLC system (Agilent Technologies, Palo Alto, CA, USA), coupled with a Waters Acquity HSS T3 (2.1 × 100 mm × 1.8 um) separation column (Waters, Milford, MA, USA) and a HSS T3 (2.1 × 5 mm × 1.8 um) guard column, was used to obtain the chromatographic separations with the column temperature set at 40 °C. The sample injection volume was 5 µL. The mobile phases consisted of solvents A (0.1% formic acid in ddH$_2$O) and B (0.1% formic acid in acetonitrile) with a 0.45 mL/min flow rate. The gradient program was as follows: 0–1 min, 100% A; 1–16 min, a linear gradient to 70% A; 16–21 min, a linear gradient to 5% A, and hold for 1.5 min; returning to 100% A over 1 min, and hold for 5 min for column re-equilibration. An Agilent 6545 quadrupole time-of-flight (Q-TOF) mass spectrometer was used for mass spectrometric analysis. The positive electrospray ionization (ESI) mode was applied for mass spectral (70–1000 m/z) data collection. The ESI capillary voltage was 3.5 kV, the drying gas flow rate was 8.0 L/min, the temperature of the nitrogen gas was 325 °C, the nebulizer gas pressure was 30 psig, the fragmentor voltage was 130 V, the skimmer was 45 V, and the Oct 1 RF Vpp was 750 V. Mass data were acquired using Agilent MassHunter B.06.

2.9.3. Metabolomics Data Processing

The raw data were preprocessed (undergoing peak detection, deconvolution, and alignment) using Agilent ProFinder (v B.08). The peak area matrix was imported to MetaboAnalyst 4.0 for principal component analysis (PCA) and hierarchical cluster analysis (HCA). Variables with missing values were dealt a small value (half of the minimum values in the original data). The data were normalized by the sum, transformed by generalized log transformation, and auto scaled. Pairwise comparisons (16 d versus 2 d, 9 d versus 2 d, and 16 d versus 9 d) were performed, respectively, and features with a minimum fold change of 2 (ratio > 2 or <0.5; FDR < 0.05) were considered to have been changed significantly in the three comparison groups. Identifications were aided by performing data-dependent MS/MS collection on composite samples with 10 eV, 20 eV, and 40 eV of collision energy, respectively. Annotation was achieved based on the database of METLIN (www.metlin.scripps.edu (accessed on 20 January 2021)), with a mass tolerance of 10 ppm for MS1 and a mass tolerance of 20 ppm for MS2.

2.10. Statistical Analysis

The experimental design of this study was a completely randomized design. Each loin was used as an experimental unit. A repeated measure design was used to detect instrumental color parameters at three aging periods of retail display. The data of meat quality attributes were analyzed using SPSS (SPAA Inc., Chicago, IL, USA) for one-way ANOVA to evaluate the significance of the main treatment (aging). Significance was defined at the level of $p < 0.05$.

3. Results and Discussion

3.1. pH

The pH of the muscles was significantly affected by aging (Table 1). At 24 h postmortem, the ultimate pH dropped to 5.65 due to the accumulation of lactate and hydrogen ions from postmortem glycolysis and ATP hydrolysis [26]. The pH declined continually to 5.55 after 2 d of aging, and increased ($p < 0.05$) to 5.59 and 5.57 after 9 d and 16 d of aging, respectively. The increased pH may be due to the changes in charge caused by proteolytic enzymes during postmortem aging [27].

Table 1. The pH and water-holding capacity changes of pork loins during postmortem aging.

Aging Time	1 d (24 h)	2 d	9 d	16 d
pH	5.65 ± 0.01 [a]	5.55 ± 0.01 [c]	5.59 ± 0.01 [b]	5.57 ± 0.01 [b]
Purge loss (%)	-	3.23 ± 0.30 [c]	5.41 ± 0.50 [b]	7.80 ± 0.55 [a]
Drip loss (%)	-	6.17 ± 0.31 [a]	2.74 ± 0.16 [b]	1.46 ± 0.18 [c]
Cooking loss (%)	-	28.33 ± 0.90	28.01 ± 0.73	25.74 ± 1.06

Results are expressed as the mean ± standard error. Means with different letters ([a–c]) in a row are different ($p < 0.05$).

3.2. WHC

Purge loss increased from 3.23% at 2 d postmortem to 7.80% at 16 d postmortem ($p < 0.05$) (Table 1). The increase in purge loss with aging could be due to the degradation of proteins by proteolysis during aging, resulting in a breakdown in the myofibrils, cell membrane structure, and intracellular cytoskeleton, which induces the migration of myowater more easily from the intramyofibrilar space [5].

Drip loss dramatically decreased from 6.17% to 1.46% with aging ($p < 0.05$) (Table 1). Higher drip loss has been shown to occur during first 2 days postmortem, which corresponds to the findings from Straadt et al. [28], wherein drip loss increased from 24 h to 48 h postmortem and stayed constant until 4 d postmortem, after which point it decreased up until 14 d postmortem. From one side, the decreased drip loss should indicate the increased WHC of meat during aging according to the hypothesis of Kristensen and Purslow [29], which showed that degradations in the cytoskeleton during aging reduce or remove the linkage between the rigor-induced lateral shrinkage of myofibrils and the shrinkage of the whole muscle fiber. Therefore, muscle myofibrils are observed to "swell" and are capable of holding more water during aging [28]. However, an increased purge loss from meat during aging in a vacuum bag was still observed (Table 1), meaning that more water was exuded from meat, which may result in a lesser volume of water in the meat that was available to be released as drip loss.

As for cooking loss, no significant differences among aging periods were found, although 16 d postmortem showed a lower value numerically (Table 1). The weight loss of the meat during cooking is related to how much water is present and how easily it can leave the muscle structure network [5]. Purge loss increased following aging, indicating a lesser volume of water available in the muscle to be easily evaporated. Thus, no significantly changed cooking loss may suggest a decreased water-holding capacity as aging progresses.

3.3. Color Stability

The effect of wet-aging on the instrumental color parameters of the pork samples during display (0, 1, 2, 3, 4, and 5 days) is reported in Figure 1. Lightness (L*) was affected by aging time; 2 d aged meat showed lower values than did 9 d and 16 d aged meat at the first 3-day display. Additionally, L* development showed a significant increase ($p < 0.01$) in 2 d aged meat during display periods. Pork cuts from 16 d aged sections showed higher ($p < 0.01$) a* values than did 2 d and 9 d aged samples at the first 2-day display. However, at the end of the display period, 2 d aged cuts presented significantly higher ($p < 0.05$) a* values than did 9 d and 16 d aged ones. Moreover, a* values were all affected by display ($p < 0.01$) regardless of aging treatments and showed a significantly decreased tendency from that exhibited during day 1 display. Chroma, which represents the color intensity of the meat, was also affected by aging time ($p < 0.01$). Compared with shorter aging treatments, 16 d aged cuts showed the highest chroma at the first 2-day display. However, they presented more significant declines and showed the lowest chroma at the end of display. The decrease in chroma from 1 to 5 days of display indicates a gradually increased accumulation of surface metmyoglobin. These results may suggest that longer aging treatment could reduce meat color stability during display period [30]. Aged meat blooms more rapidly but subsequently browns earlier than fresh meat [31]. Previous research also indicated that the lower color stability of aged meat compared with unaged meat becomes evident after 3 days of display [32].

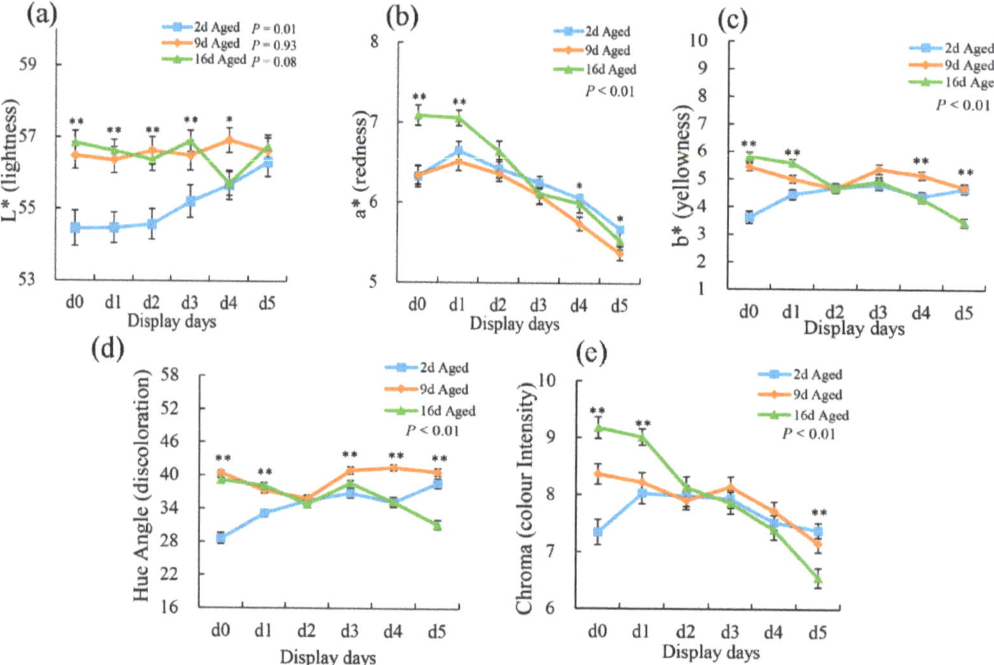

Figure 1. The effect of aging on the changes in L* (**a**), a* (**b**), b* (**c**), hue angle (**d**), and chroma (**e**) of pork steak during 5 days of display. The results are expressed as the mean ± standard error. The significant differences among three aging treatments at the same display time were marked with asterisks (* means $p < 0.05$; ** means $p < 0.01$).

3.4. Shear Force and MFI

The shear force values of the pork samples were affected ($p < 0.05$) by aging time (Figure 2a). Two-day aged samples showed the highest shear force value, and the shear force value decreased as the aging period extended, which suggested an increased tenderness of the meat. At the same time, MFI significantly increased ($p < 0.05$) from 80.92 after 2 d postmortem aging to 110.76 with 9 d aging, and no significant difference was found between 9 d and 16 d aging treatments (Figure 2b). A higher MFI indicates a more severe myofibrillar degradation and thus a subsequently more tender meat [33]. Aging improves the tenderness of the meat through disruption of the muscle ultrastructure by intracellular proteolytic systems (including calpains, the lysosomal proteases, and cathepsins), which have the capability of degrading myofibrillar and cytoskeletal proteins [34].

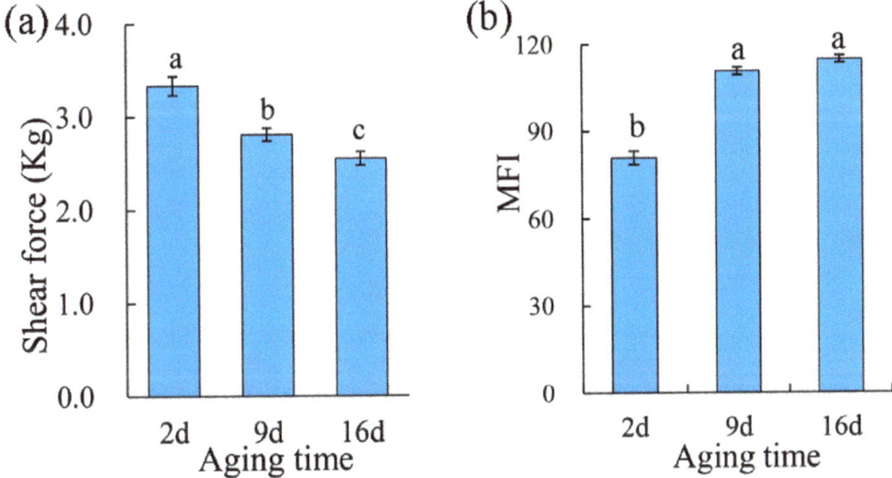

Figure 2. The effect of aging treatments on shear force (**a**) and the myofibril fragmentation index (**b**) of the pork meat. Results are expressed as the mean ± standard error. Means with different letters (a–c) are different ($p < 0.05$).

3.5. Oxidative Stability

TBARS was performed to evaluate the lipid oxidation of the pork meat during aging. As showed in Figure 3a, there was a significant increase ($p < 0.05$) in the TBARS values of pork samples after 9 d aging than that of 2 d aged samples; however, no significant difference was found between 9 d and 16 d aging treatments. As for protein oxidation, carbonyl content increased numerically during aging even though no statistical significance was found (Figure 3b). Lipid oxidation is a critical deterioration reaction in meat and is positively correlated with oxymyoglobin oxidation [35]. During postmortem aging, muscle cells gradually lose their ability to maintain reducing conditions due to the collapse of the antioxidant system and structural integrity, which may lead postmortem muscles to be more susceptible to oxidation. Similarly, 14-day vacuum aged beef muscle indicated a higher lipid oxidation than that in nonvacuum packaged meat, as described by Popova et al. [36]. However, some researchers reported no aging effect on the initial TBARS from unaged or wet aged beef for 14 days [37] or no significant change in the TBARS of beef during aging for 14 days [38].

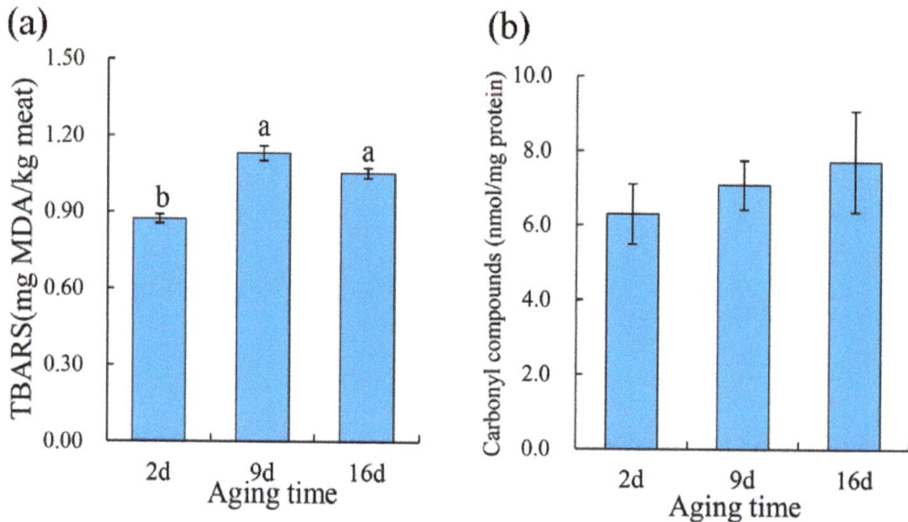

Figure 3. The effect of aging treatments on lipid (**a**) and protein (**b**) oxidation of pork meat. Results are expressed as the mean ± standard error. Means with different letters (a, b) are different (*p* < 0.05).

3.6. Metabolomics Profiling of Pork Exudate

A total of 1483 features were detected via UHPLC-QTOF-MS in the 15 purge samples (5 biological replicates × 3 aging treatments), and 20 of them were deleted due to having more than 50% of their values missing. Thus, a total of 1463 features were used for PCA and HCA analyses. There were complete separations among three different aging treatments from the PCA plot (Figure 4a), which could indicate very distinct purge metabolome profiles observed with different aging days. Similarly, the heatmap also showed two distinct clusters in which 2 d purge samples aggregated together and significantly differed from 9 d and 16 d samples (Figure 4b). Although 9 d and 16 d samples presented similar metabolome profiles, these samples still produced some distinct clusters, respectively. The purge metabolome from loin 9 showed some differences from the other metabolomes, which could be due to differences in the biological samples themselves. Basically, since the metabolome profiles of pork purge were significantly affected by aging treatments, it is worthwhile to emphasize that purge from shorter aged (2 d postmortem aged) pork presented totally different metabolome profiles than did longer aged pork purge (9 d and 16 d).

A total of 933 features changed significantly in three pairwise comparisons (16 d vs. 2 d, 9 d vs. 2 d, and 16 d vs. 9 d) (Figure 4c) with a fold change of 2 (ratio > 2 or <0.5) and a false discovery rate (FDR) < 0.05, in which 34 features were tentatively identified via matching METLIN database according to MS/MS information (Table 2).

Ten compounds were overabundant in 2 d postmortem aged purge (Table 2), and these compounds were involved in ATP synthesis and metabolism (creatine, hypoxanthine), histidine metabolism (L-Histidine, carnosine), phenylalanine metabolism (2-Phenylacetamide), and the antioxidation of reactive oxygen species (ROS) and free radicals (glutathione, oxidized GSSH). Postmortem ATP synthesis and metabolism radically determines most of the critical qualities of raw meat. Creatine has the ability to increase muscle stores of phosphocreatine, potentially increasing the muscle's ability to resynthesize ATP from ADP in order to meet increased energy demands [39]. The muscle ATP concentration remains stable through the phosphocreatine shuttle early in the postmortem process. However, muscle stores of phosphocreatine are also limited (around 25 mmol/g of muscle tissue) and the phosphocreatine buffer system can only sustain postmortem cellular ATP for a brief period [26], even though creatine and phosphocreatine could be still detected in certain concentrations in pork meat after 44 d of aging [40]. The decline of creatine content in pork purge during aging in the present study coincided with the decreased tendency of creatine and phosphocreatine levels in pork meat after longer aging treatments [40]. This might suggest a progressive collapse of the phosphocreatine shuttle during aging. Meanwhile, hypoxanthine (a degradation product of ATP) also declined in pork purge during aging. It was reported that the ATP content will rapidly decrease and almost be exhausted at 24 h postmortem in pork longissimus lumborum (LL) muscles, followed by an increase in the downstream product of ATP degradation [41]. However, hypoxanthine can be oxidized into xanthine via xanthine dehydrogenase; thus, it could be decreased in pork purge alongside aging.

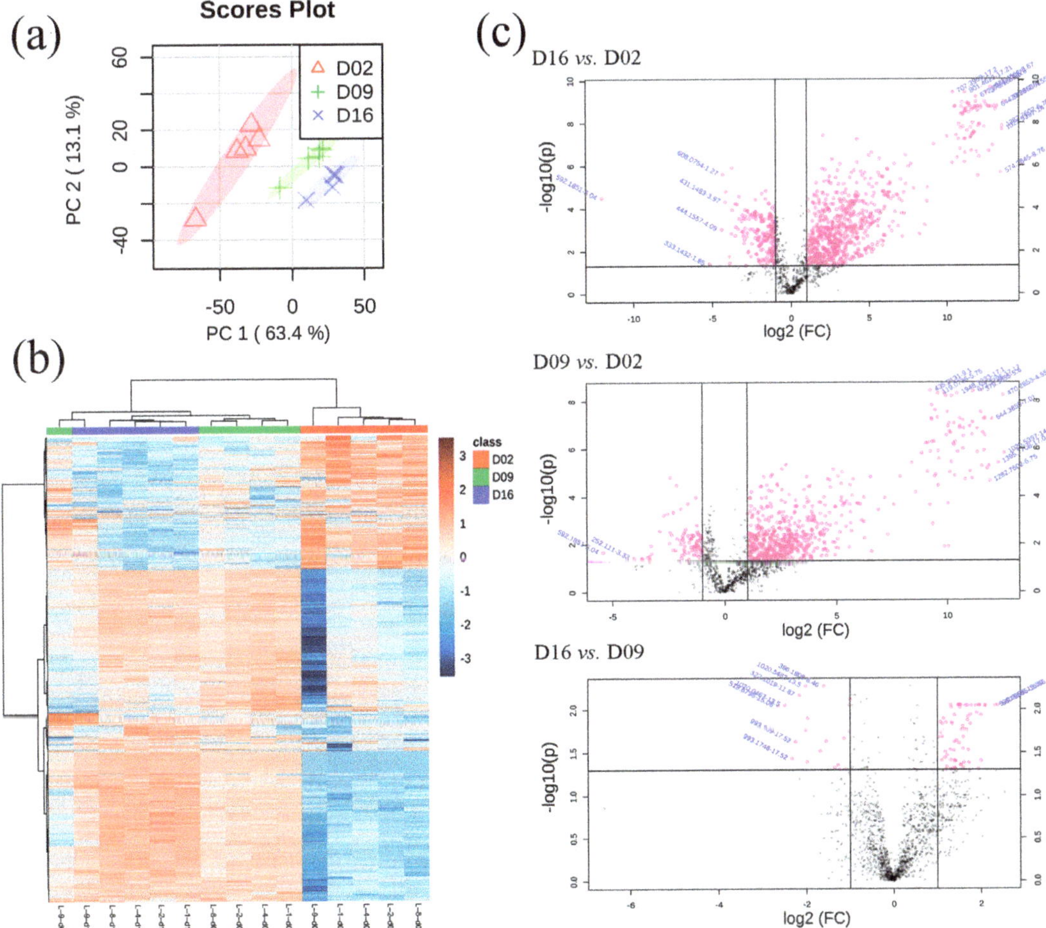

Figure 4. The principal component analysis (PCA) score plots (**a**), hierarchical cluster analysis (HCA) score plots (**b**), and volcano plots (**c**) of metabolome in pork exudate. D02, D09, and D16 refer to pork exudate from 2 d, 9 d, and 16 d postmortem aging treatments, respectively.

When animals are slaughtered for meat, oxygen deficiency results in the impaired mitochondrial respiration, producing ROS [40]. Glutathione and its redox forms (oxidized glutathione as GSSH, reduced glutathione as GHS) plays a critical role in preventing damage caused by ROS (such as free radicals, peroxides, lipid peroxides, and heavy metals) [42] for cellular components. GSSG is produced when peroxides are detoxified by GSH peroxidase and are recycled back to their reduced form by GSH reductase, at the expense of nicotinamide adenine dinucleotide phosphate (NADPH). The significantly lower level of GSSG in 16 d aged purge samples as compared to that in 2 d aged samples might indicate the gradually incapacitated antioxidant defense system in pork muscle during the aging period. Meanwhile, carnosine and L-histidine showed significant lower levels in 16 d aged purge compared to those in 2 d aged samples. Carnosine, a dipeptide composed of L-histidine and β-alanine, is a potential endogenous antioxidant that scavenges ROS as well as inhibits lipid oxidation during oxidative stress [43–45]. L-histidine has also been demonstrated to have an antioxidative action and has proved to be an excellent scavenger of reactive $^1O_2$ species [46]. Taken together, the decrease in GSSG carnosine and L-histidine identified in pork purge during aging would be likely associated with increased lipid and protein oxidation (Figure 3) in pork meat.

Table 2. Tentatively identified metabolites that have changed significantly among the three aging treatments.

Name	Mass	*m/z*	MS1 Error (ppm)	RT [1]	Adduct Ion Name	D16 vs. D02		D09 vs. D02		D16 vs. D09	
						Fold-Change	FDR [2]	Fold-Change	FDR	Fold-Change	FDR
Compounds overabundant in 2 d postmortem aging (12)											
Creatine	131.0703	132.0774	−5.15	0.92	(M+H)⁺	0.457	0.026	0.372	0.004		
2-Phenylacetamide	135.0795	136.0764	−5.08	3.13	(M+H)⁺	0.344	0.019	0.339	0.006		
Hypoxanthine	136.0382	137.0466	−5.75	1.32	(M+H)⁺	0.127	0.004	0.238	0.048		
N-Methylisoleucine	145.1103	146.1183	−4.81	4.31	(M+H)⁺	0.426	0.001				
L-Histidine	155.0698	156.0773	−3.2	1.04	(M+H)⁺	0.473	0.000				
Carnosine	226.1071	227.1149	−4.69	1.03	(M+H)⁺	0.420	0.011				
Isobutyryl carnitine	232.1432	232.1559	−6.88	4.75	(M)⁺	0.423	0.010				
His Glu	284.1101	285.1208	−5.29	1.36	(M+H)⁺	0.413	0.000				
Lys Pro Ile	356.2432	357.2509	−3.5	0.83	(M+H)⁺	0.492	0.002				
Glutathione, oxidized	612.1505	613.1613	−3.39	1.49	(M+H)⁺	0.432	0.000				
Compounds overabundant in 9 d and 16 d postmortem aging (27)											
Pro Ala	186.1004	187.1086	−4.88	1.12	(M+H)⁺	3.170	0.001	2.206	0.006		
Ala Val	188.1155	189.1242	−4.23	1.32	(M+H)⁺	2.175	0.000				
Ile Gly	188.1162	189.1242	−4.23	2.86	(M+H)⁺	3.190	0.000	2.198	0.000		
L-Leucyl-L- Alanine	202.1319	203.14	−4.93	2.53	(M+H)⁺	3.316	0.003	3.011	0.009		
Spermine	202.1317	203.2239	−4.12	3.28	(M+H)⁺	2.899	0.018	2.480	0.042		
Val Ser	204.111	205.1196	−6.27	0.88	(M+H)⁺	2.766	0.001	2.018	0.011		
Pro Thr	216.1095	217.1191	−3.88	1.25	(M+H)⁺	4.626	0.001	3.379	0.004		
Val Val	216.1477	217.1556	−4.14	3.43	(M+H)⁺	7.910	0.000	5.018	0.000		
Ile Ser	218.1267	219.1347	−3.7	1.2	(M+H)⁺	4.945	0.000	3.137	0.002		
Val Leu	230.1631	231.1711	−3.57	5.69	(M+H)⁺	4.716	0.000	3.693	0.003		
Asn Val	231.1215	232.1303	−4.81	1.3	(M+H)⁺	4.741	0.000	4.416	0.012		
Leu Thr	232.1424	233.1506	−4.45	1.9	(M+H)⁺	5.627	0.003	3.262	0.042		
Ala Phe	236.1162	237.1244	−4.24	5.76	(M+H)⁺	8.885	0.000	4.650	0.007		
Ile Ile	244.1783	245.1869	−3.77	7.8	(M+H)⁺	6.780	0.001	3.081	0.014		
Ile Asp	246.1218	247.1296	−3.2	4.05	(M+H)⁺	2.105	0.000	2.146	0.000		
Asp Leu	246.122	247.1296	−3.2	4.43	(M+H)⁺	5.903	0.000	4.222	0.000		
Tyr Ala	252.1115	253.1196	−5.21	2.42	(M+H)⁺	2.381	0.026	2.911	0.027		
Ala Ile Gly	259.1553	260.1617	−4.61	1.36	(M+H)⁺	6.617	0.000	5.237	0.000		
Pro Leu Gly	285.1685	286.1771	−3.26	6.11	(M+H)⁺	3.141	0.007	2.213	0.025		
Val Ile Gly	287.186	288.1939	−7.4	4.32	(M+H)⁺	9.503	0.000	4.396	0.007		
His Phe	302.1376	303.1463	−3.65	2.74	(M+H)⁺	9.624	0.000	5.950	0.001		
Gly Lys Leu	316.2111	317.2195	−3.54	2	(M+H)⁺	123.990	0.000	37.504	0.001	3.306	0.041
Val Val Asp	331.1744	332.1832	−4.66	3.51	(M+H)⁺	47.939	0.000	19.345	0.001		
Val Glu Glu	375.1641	376.173	−4.06	3.07	(M+H)⁺	3.433	0.005	3.348	0.024		

[1] RT is the average retention time. [2] FDR is the false discovery rate of the T-test. D02, D09, and D16 refer to pork exudate from 2 d, 9 d, and 16 d postmortem aging treatments, respectively.

At the same time, 24 compounds were identified in overabundance in 9 d and 16 d aged pork purge (longer aging treatments). Basically, most of these compounds were dipeptides and tripeptides. The significantly increased levels of these peptides in purge after 9 d of aging can be explained by accumulated proteolysis, which starts after slaughter. The postmortem proteolysis by several endogenous proteolytic enzyme systems (calpains, cathepsins, and proteasomes) is mainly responsible for the degradation of structural proteins [47], which are broken down into polypeptide fragments, inducing an increased tenderness of the meat (reflected by a declined shear force and an increased MFI, as seen in Figure 2). This is followed by peptidyl peptidases acting to generate small peptides. The aforementioned changes induce a relative significantly increased concentration of peptides in the muscle with a longer postmortem aging time [48,49]. Thus, it is speculated that increasing the amount of peptides effused from muscle with water loss and detecting them in purge as aging indicators (Table 2), could make them useful as potential biomarkers for predicting meat tenderness during aging.

4. Conclusions

In summary, the present study indicated that aging significantly affected meat quality attributes as well as meat exudate metabolome profiles. Metabolites associated with ATP synthesis and metabolism (creatine and hypoxanthine), antioxidation (GSSG and carnosine), and proteolysis (dipeptides and tripeptides) could act as potential biomarkers to monitor aging times and indicate meat quality changes, such as increased lipid and protein oxidation, discoloration, and tenderness during aging processing. Further studies regarding target compound (potential biomarkers) quantification and proteome profiles of meat exudate would be highly warranted to provide more valuable information associated with meat quality.

Author Contributions: Conceptualization, Q.Y. and Y.H.B.K.; methodology, Q.Y. and Y.H.B.K.; investigation, Q.Y. and Y.H.B.K.; resources, Y.H.B.K.; data curation, Q.Y., B.C., and T.S.; writing—original draft preparation, Q.Y.; writing—review and editing, Y.H.B.K.; visualization, Q.Y.; supervision, Y.H.B.K.; project administration, Y.H.B.K.; funding acquisition, Y.H.B.K. All authors have read and agreed to the published version of the manuscript.

Funding: This activity was funded by Purdue University as part of AgSEED Crossroads funding to support Indiana's Agriculture and Rural Development.

Acknowledgments: The authors would like to acknowledge all members of the Meat Science and Muscle Biology Laboratory for the successful completion of the sample and data collection.

Conflicts of Interest: The authors declare no competing financial interest with any organization regarding this manuscript.

References

1. Kim, Y.H.B.; Ma, D.; Setyabrata, D.; Farouk, M.M.; Lonergan, S.M.; Huff-Lonergan, E.; Hunt, M.C. Understanding Postmortem Biochemical Processes and Post-Harvest Aging Factors to Develop Novel Smart-Aging Strategies. *Meat Sci.* **2018**, *144*, 74–90. [CrossRef] [PubMed]
2. Mcmillin, K.W. Advancements in Meat Packaging. *Meat Sci.* **2017**, *132*, 153–162. [CrossRef]
3. Colle, M.J.; Richard, R.P.; Killinger, K.M.; Bohlscheid, J.C.; Gray, A.R.; Loucks, W.I.; Day, R.N.; Cochran, A.S.; Nasados, J.A.; Doumit, M.E. Influence of Extended Aging on Beef Quality Characteristics and Sensory Perception of Steaks from the Gluteus Medius and Longissimus Lumborum. *Meat Sci.* **2015**, *110*, 32–39. [CrossRef] [PubMed]
4. Hodges, J.H.; Cahill, V.R.; Ockerman, H.W. Effect of Vacuum Packaging on Weight Loss, Microbial Growth and Palatability of Fresh Beef Wholesale Cuts. *J. Food Sci.* **1974**, *39*, 143–146. [CrossRef]
5. Warner, R.D. The Eating Quality of Meat—Iv Water-Holding Capacity and Juiciness. In *Lawrie's Meat Science*, 8th ed.; Toldra', F., Ed.; Woodhead Publishing: Cambridge, UK, 2017; Chapter 14; pp. 419–459.
6. Gariepy, C.; Amiot, J.; Simard, R.E.; Boudreau, A.; Raymond, D.P. Effect of Vacuum Packing and Storage in Nitrogen and Carbon Dioxide Atmospheres on the Quality of Fresh Rabbit Meat. *J. Food Qual.* **1986**, *9*, 289–309. [CrossRef]
7. Castejón, D.; García-Segura, J.M.; Escudero, R.; Herrera, A.; Cambero, M.I. Metabolomics of Meat Exudate: Its Potential to Evaluate Beef Meat Conservation and Aging. *Anal. Chim. Acta* **2015**, *901*, 1–11. [CrossRef]
8. Savage, A.W.J.; Warriss, P.D.; Jolley, P.D. The Amount and Composition of the Proteins in Drip from Stored Pig Meat. *Meat Sci.* **1990**, *27*, 289–303. [CrossRef]
9. Huff-Lonergan, E.; Lonergan, S.M. Mechanisms of Water-Holding Capacity of Meat: The Role of Postmortem Biochemical and Structural Changes. *Meat Sci.* **2005**, *71*, 194–204. [CrossRef]
10. Kim, G.D.; Jung, E.Y.; Lim, H.J.; Yang, H.S.; Joo, S.T.; Jeong, J.Y. Influence of Meat Exudates on the Quality Characteristics of Fresh and Freeze-Thawed Pork. *Meat Sci.* **2013**, *95*, 323–329. [CrossRef]
11. Traore, S.; Aubry, L.; Gatellier, P.; Przybylski, W.; Jaworska, D.; Kajak-Siemaszko, K.; Santé-Lhoutellier, V. Higher Drip Loss Is Associated with Protein Oxidation. *Meat Sci.* **2012**, *90*, 917–924. [CrossRef]
12. García-García, A.B.; Herrera, A.; Fernández-Valle, M.E.; Cambero, M.I.; Castejón, D. Evaluation of E-Beam Irradiation and Storage Time in Pork Exudates Using Nmr Metabolomics. *Food Res. Int.* **2019**, *120*, 553–559. [CrossRef]
13. Xing, T.; Zhao, X.; Xu, X.; Li, J.; Zhang, L.; Gao, F. Physiochemical Properties, Protein and Metabolite Profiles of Muscle Exudate of Chicken Meat Affected by Wooden Breast Myopathy. *Food Chem.* **2020**, *316*, 126271. [CrossRef]

14. Hu, C.; Xu, G. Mass-Spectrometry-Based Metabolomics Analysis for Foodomics. *TrAC Trends Anal. Chem.* **2013**, *52*, 36–46. [CrossRef]

15. Abraham, A.; Dillwith, J.W.; Mafi, G.G.; Vanoverbeke, D.L.; Ramanathan, R. Metabolite Profile Differences between Beef Longissimus and Psoas Muscles During Display. *Meat Muscle Biol.* **2017**, *1*, 18–27. [CrossRef]

16. Ma, D.; Kim, Y.H.B.; Cooper, B.; Oh, J.-H.; Chun, H.; Choe, J.-H.; Schoonmaker, J.P.; Ajuwon, K.; Min, B. Metabolomics Profiling to Determine the Effect of Postmortem Aging on Color and Lipid Oxidative Stabilities of Different Bovine Muscles. *J. Agric. Food Chem.* **2017**, *65*, 6708–6716. [CrossRef]

17. Subbaraj, A.K.; Kim, Y.H.B.; Fraser, K.; Farouk, M.M. A Hydrophilic Interaction Liquid Chromatography–Mass Spectrometry (Hilic–Ms) Based Metabolomics Study on Colour Stability of Ovine Meat. *Meat Sci.* **2016**, *117*, 163–172. [CrossRef] [PubMed]

18. Yu, Q.; Tian, X.; Shao, L.; Li, X.; Dai, R. Targeted Metabolomics to Reveal Muscle-Specific Energy Metabolism between Bovine Longissimus Lumborum and Psoas Major During Early Postmortem Periods. *Meat Sci.* **2019**, *156*, 166–173. [CrossRef]

19. Setyabrata, D.; Ma, D.; Cooper, B.R.; Sobreira, T.J.; Kim, Y.H.B. Metabolomics Profiling of Meat Exudate to Understand the Impact of Postmortem Aging on Oxidative Stability of Beef Muscles. In Proceedings of the ICoMST-Proceedings, Melbourne, Australia, Melbourne, Australia, 12–17 August 2018.

20. Setyabrata, D.; Kim, Y.H.B. Impacts of Aging/Freezing Sequence on Microstructure, Protein Degradation and Physico-Chemical Properties of Beef Muscles. *Meat Sci.* **2019**, *151*, 64–74. [CrossRef] [PubMed]

21. Kim, Y.H.B.; Meyers, B.; Kim, H.-W.; Liceaga, A.M.; Lemenager, R.P. Effects of Stepwise Dry/Wet-Aging and Freezing on Meat Quality of Beef Loins. *Meat Sci.* **2017**, *123*, 57–63. [CrossRef]

22. *AMSA Meat Color Measurement Guidelines*; American Meat Science Association: Champaign, IL, USA, 2012; pp. 1–17.

23. Culler, R.; Smith, G.; Cross, H. Relationship of Myofibril Fragmentation Index to Certain Chemical, Physical and Sensory Characteristics of Bovine Longissimus Muscle. *J. Food Sci.* **1978**, *43*, 1177–1180. [CrossRef]

24. Vossen, E.; De Smet, S. Protein Oxidation and Protein Nitration Influenced by Sodium Nitrite in Two Different Meat Model Systems. *J. Agric. Food Chem.* **2015**, *63*, 2550–2556. [CrossRef] [PubMed]

25. Bligh, E.G.; Dyer, W.J. A Rapid Method of Total Lipid Extraction and Purification. *Can. J. Biochem. Physiol.* **1959**, *37*, 911–917. [CrossRef] [PubMed]

26. Matarneh, S.K.; England, E.M.; Scheffler, T.L.; Gerrard, D.E. The Conversion of Muscle to Meat. In *Lawrie's Meat Science*, 8th ed.; Toldra', F., Ed.; Woodhead Publishing: Cambridge, UK, 2017; Chapter 5, pp. 159–185.

27. Boakye, K.; Mittal, G. Changes in Ph and Water Holding Properties of Longissimus Dorsi Muscle During Beef Ageing. *Meat Sci.* **1993**, *34*, 335–349. [CrossRef]

28. Straadt, I.K.; Rasmussen, M.; Andersen, H.J.; Bertram, H.C. Aging-Induced Changes in Microstructure and Water Distribution in Fresh and Cooked Pork in Relation to Water-Holding Capacity and Cooking Loss—A Combined Confocal Laser Scanning Microscopy (Clsm) and Low-Field Nuclear Magnetic Resonance Relaxation Study. *Meat Sci.* **2007**, *75*, 687–695. [CrossRef] [PubMed]

29. Kristensen, L.; Purslow, P.P. The Effect of Ageing on the Water-Holding Capacity of Pork: Role of Cytoskeletal Proteins. *Meat Sci.* **2001**, *58*, 17–23. [CrossRef]

30. Kim, Y.H.B.; Frandsen, M.; Rosenvold, K. Effect of Ageing Prior to Freezing on Colour Stability of Ovine Longissimus Muscle. *Meat Sci.* **2011**, *88*, 332–337. [CrossRef]

31. Johnson, D.E.; Knight, M.K.; Ledward, D.A. Colour of Raw and Cooked Meat. In *The Chemistry of Muscle-Based Foods*; The Royal Society of Chemistry: Cambridge, UK, 1992; pp. 128–144.

32. Vitale, M.; Pérez-Juan, M.; Lloret, E.; Arnau, J.; Realini, C.E. Effect of Aging Time in Vacuum on Tenderness, and Color and Lipid Stability of Beef from Mature Cows During Display in High Oxygen Atmosphere Package. *Meat Sci.* **2014**, *96*, 270–277. [CrossRef]

33. Marino, R.; Albenzio, M.; Della Malva, A.; Santillo, A.; Loizzo, P.; Sevi, A. Proteolytic Pattern of Myofibrillar Protein and Meat Tenderness as Affected by Breed and Aging Time. *Meat Sci.* **2013**, *95*, 281–287. [CrossRef]

34. Bhat, Z.F.; Morton, J.D.; Mason, S.L.; Bekhit, A.E.-D.A. Role of Calpain System in Meat Tenderness: A Review. *Food Sci. Hum. Wellness* **2018**, *7*, 196–204. [CrossRef]

35. Chan, W.K.; Faustman, C.; Decker, E.A. Oxymyoglobin Oxidation as Affected by Oxidation Products of Phosphatidylcholine Liposomes. *J. Food Sci.* **1997**, *62*, 709–712. [CrossRef]

36. Popova, T.; Marinova, P.; Vasileva, V.; Gorinov, Y.; Lidji, K. Oxidative Changes in Lipids and Proteins in Beef During Storage. *Arch. Zootech.* **2009**, *12*, 30–38.

37. Jiang, T.; Busboom, J.; Nelson, M.; O'fallon, J.; Ringkob, T.; Rogers-Klette, K.; Joos, D.; Piper, K. The Influence of Forage Diets and Aging on Beef Palatability. *Meat Sci.* **2010**, *86*, 642–650. [CrossRef]

38. Spanier, A.; Flores, M.; Mcmillin, K.; Bidner, T. The Effect of Post-Mortem Aging on Meat Flavor Quality in Brangus Beef. Correlation of Treatments, Sensory, Instrumental and Chemical Descriptors. *Food Chem.* **1997**, *59*, 531–538. [CrossRef]

39. Wallimann, T.; Wyss, M.; Brdiczka, D.; Nicolay, K.; Eppenberger, H. Intracellular Compartmentation, Structure and Function of Creatine Kinase Isoenzymes in Tissues with High and Fluctuating Energy Demands: The 'phosphocreatine Circuit' for Cellular Energy Homeostasis. *Biochem. J.* **1992**, *281*, 21. [CrossRef]

40. Lana, A.; Longo, V.; Dalmasso, A.; D'alessandro, A.; Bottero, M.T.; Zolla, L. Omics Integrating Physical Techniques: Aged Piedmontese Meat Analysis. *Food Chem.* **2015**, *172*, 731–741. [CrossRef] [PubMed]

41. Muroya, S.; Oe, M.; Nakajima, I.; Ojima, K.; Chikuni, K. Ce-Tof Ms-Based Metabolomic Profiling Revealed Characteristic Metabolic Pathways in Postmortem Porcine Fast and Slow Type Muscles. *Meat Sci.* **2014**, *98*, 726–735. [CrossRef]

42. Pompella, A.; Visvikis, A.; Paolicchi, A.; De Tata, V.; Casini, A.F. The Changing Faces of Glutathione, a Cellular Protagonist. *Biochem. Pharmacol.* **2003**, *66*, 1499–1503. [CrossRef]

43. Babizhayev, M.A.; Seguin, M.; Gueyne, J.; Evstigneeva, R.; Ageyeva, E.; Zheltukhina, G. L-Carnosine (B-Alanyl-L-Histidine) and Carcinine (B-Alanylhistamine) Act as Natural Antioxidants with Hydroxyl-Radical-Scavenging and Lipid-Peroxidase Activities. *Biochem. J.* **1994**, *304*, 509–516. [CrossRef]

44. Nagasawa, T.; Yonekura, T.; Nishizawa, N.; Kitts, D.D. In Vitro and in Vivo Inhibition of Muscle Lipid and Protein Oxidation by Carnosine. *Mol. Cell. Biochem.* **2001**, *225*, 29–34. [CrossRef] [PubMed]

45. Zhou, S.; Decker, E.A. Ability of Carnosine and Other Skeletal Muscle Components to Quench Unsaturated Aldehydic Lipid Oxidation Products. *J. Agric. Food Chem.* **1999**, *47*, 51–55. [CrossRef]

46. Obata, T.; Inada, T. Protective Effect of Histidine on Mpp+-Induced Hydroxyl Radical Generation in Rat Striatum. *Brain Res.* **1999**, *817*, 206–208. [CrossRef]

47. Lana, A.; Zolla, L. Proteolysis in Meat Tenderization from the Point of View of Each Single Protein: A Proteomic Perspective. *J. Proteom.* **2016**, *147*, 85–97. [CrossRef] [PubMed]

48. Claeys, E.; De Smet, S.; Balcaen, A.; Raes, K.; Demeyer, D. Quantification of Fresh Meat Peptides by Sds–Page in Relation to Ageing Time and Taste Intensity. *Meat Sci.* **2004**, *67*, 281–288. [CrossRef] [PubMed]

49. Fu, Y.; Young, J.F.; Therkildsen, M. Bioactive Peptides in Beef: Endogenous Generation through Postmortem Aging. *Meat Sci.* **2017**, *123*, 134–142. [CrossRef] [PubMed]

MDPI

St. Alban-Anlage 66

4052 Basel

Switzerland

Tel. +41 61 683 77 34

Fax +41 61 302 89 18

www.mdpi.com

Foods Editorial Office

E-mail: foods@mdpi.com

www.mdpi.com/journal/foods